Praise for

Seed Libraries

Seed Libraries is must-read for anyone concerned about the absorption of small, regional seed companies into large, petrochemical multinationals; the rise of GMO seeds; and the loss of genetic diversity in our food crops. Cindy Conner introduces a movement to keep seeds in the hands of the people while revitalizing public libraries and communities. She encourages us to set up our own local seed libraries with step-by-step instructions on getting started, as well as how to keep it going. As Cindy says, "whoever controls the seeds controls the food supply."

— Ira Wallace, Southern Exposure Seed Exchange and author,
The Timber Press Guide to Vegetable Gardening in the Southeast

Cindy Conner is a woman after my own heart: focused on cooperating with life's generosity to shape a bountiful future. *Seed Libraries* opens wide a door to the world of seed saving for the curious and committed alike. The only seeds I now "save" are from generations of neglected self-seeding Russian Kale plants, but with this book I believe I could be my own empowered seed-mistress.

—Vicki Robin, author, *Blessing the Hands that Feed Us*

Seed saving groups of all kinds are sprouting up across the country. *Seed Libraries* adds important perspective to the resources available for gardeners to develop their own living, resilient, and fundamentally nurturing seed sources. Rather than saying there is just one way to save and share seeds, this book will help groups find their own path and reassures us that diversity is a cultural and genetic necessity for both seeds and healthy communities.

— Ken Greene, Founder, Hudson Valley Seed Library,
and board member, Organic Seed Alliance.

As someone who has been involved with seeds for the past 30 years, I highly recommend Cindy Conner's book, *Seed Libraries*. She has thoroughly researched her subject and has written a full and fascinating account of the seed movement in North America. Not only does her book have much great advice and tips about starting seed libraries, her friendly writing style makes the book such a pleasure to read. What warmed my heart most of all were Cindy's many poetic gems about seeds, observations that could only come after long and passionate intimacy with them. Way to go, Cindy!

— Dan Jason, President, Salt Spring Sanctuary Society
and owner, Salt Spring Seeds

Cindy Conner brilliantly chronicles the seed library movement and provides practical tools and strategies on how to preserve our genetic and cultural heritage in seed libraries.

— Rebecca Newburn, Co-Founder,
Richmond Grows Seed Lending Library

Conner's informative book, *Seed Libraries*, is a must-read for anyone embarking on the task of setting up their own seed library, or those just interested in becoming more informed on the issue of genetic diversity in our food systems. It combines practical knowledge with the philosophy behind seed libraries and would be useful in your first or tenth year of operating a seed library and saving seeds. Highly recommended!

— Paul Hrycyk, Seed Library Coordinator,
Seeds of Diversity

Seed
Libraries

and other means of
keeping seeds *in the*
HANDS *of the*
PEOPLE

CINDY CONNER

new society
PUBLISHERS

Books for Wiser Living

recommended by *Mother Earth News*

T ODAY, MORE THAN EVER BEFORE, our society is seeking ways to live more conscientiously. To help bring you the very best inspiration and information about greener, more sustainable lifestyles, *Mother Earth News* is recommending select books from New Society Publishers. For more than 30 years, *Mother Earth News* has been North America's "Original Guide to Living Wisely," creating books and magazines for people with a passion for self-reliance and a desire to live in harmony with nature. Across the countryside and in our cities, New Society Publishers and *Mother Earth News* are leading the way to a wiser, more sustainable world. For more information, please visit MotherEarthNews.com

Cover design by Diane McIntosh.
Unless otherwise noted, images copyright © Cindy Conner.
Chapter illustration: Bean Sprout © MJ Jessen

Printed in Canada. First printing December 2014.

New Society Publishers acknowledges the financial support of the Government of Canada through the Canada Book Fund (CBF) for our publishing activities.

Paperback ISBN: 978-0-86571-782-4

eISBN: 978-1-55092-575-3

Inquiries regarding requests to reprint all or part of *Seed Libraries* should be addressed to New Society Publishers at the address below. To order directly from the publishers, please call toll-free (North America) 1-800-567-6772, or order online at www.newsociety.com

Any other inquiries can be directed by mail to:

New Society Publishers
P.O. Box 189, Gabriola Island, BC V0R 1X0, Canada
(250) 247-9737

New Society Publishers' mission is to publish books that contribute in fundamental ways to building an ecologically sustainable and just society, and to do so with the least possible impact on the environment, in a manner that models this vision. We are committed to doing this not just through education, but through action. The interior pages of our bound books are printed on Forest Stewardship Council®-registered acid-free paper that is **100% post-consumer recycled** (100% old growth forest-free), processed chlorine-free, and printed with vegetable-based, low-VOC inks, with covers produced using FSC®-registered stock. New Society also works to reduce its carbon footprint, and purchases carbon offsets based on an annual audit to ensure a carbon neutral footprint. For further information, or to browse our full list of books and purchase securely, visit our website at: www.newsociety.com

LIBRARY AND ARCHIVES CANADA CATALOGUING IN PUBLICATION

Conner, Cindy, author
 Seed libraries : and other means of keeping seeds in the hands of the people / Cindy Conner.

Includes bibliographical references and index.
Issued in print and electronic formats.
ISBN 978-0-86571-782-4 (pbk.).--ISBN 978-1-55092-575-3 (ebook)

 1. Seeds--Collection and preservation. I. Title.

SB118.38.C65 2015   631.5'21   C2014-907287-2
                  C2014-907420-4

Contents

Foreword

BILL McDORMAN AND BELLE STARR

A T ONE TIME OR ANOTHER, many of us have been pulled into the magic of seeds and have found ourselves sharing their stories. Our seed stories are as diverse and colorful as the tiny seeds themselves. Perhaps your most memorable account took place in an alley, stumbling upon an unlikely bouquet of gorgeous wildflowers cozied up next to their seed heads. A handful went into a pocket immediately. Or maybe it was one morning, while walking in your garden, you were suddenly struck by the beauty of a patch of cosmos flowers finishing off the season with their profusions of long, straight, and thin seeds ready for picking.

Our seed stories are at once universal and intensely personal. Jeanette Hart-Mann and Chrissy Orr of the Seed Broadcast travel the country in a renovated bread truck recording stories from people of all walks of life who have been touched by their experiences with seeds. Their candid, and often tearful anecdotes illuminate the power of what many have thought to be a dying art: the practice of seed saving and connecting the practice with the stories that bring seed saving to life.

Bill has many remarkable seed tales from his thirty years as a gardener and seed saver. He shares them often during the weeklong Seed

School course we created. In 1989, looking for rare heirloom tomatoes, he traveled behind the Iron Curtain to a dacha village where a seed exchange was the center of their end-of-summer activities with great accolades (and demand) going to the seed saver of the best tasting vegetables.

For the rural villagers hunkered down in Russia, seed saving was as much a means of survival as it was a cultural celebration. The time-honored rituals and necessity of saving seeds have been the very pillars of life for countless cultures over the last 12,000 years of agricultural societies — including, until recently, here in the United States. The erosion of our seed traditions (along with their ever-important stories) has edged our culture into its own crisis, though many still fail to recognize it.

Seed Libraries, and Other Means of Keeping Seeds in the Hands of the People comes to us in the nick of time. It chronicles the ascent of our reawakening to seed saving and the reasons so many have been called to return to this ritual: for its sheer practicality, its simplicity, and its critical importance. Cindy Conner brings clarity to a subject that has long been muddled by misinformation and the privatization of our seeds. It is amazing to recall that a little more than a century ago our seed heritage was collectively held as part of the public trust. Prior to the establishment of the commercial seed trade in the late 1920s, the US government freely distributed over 1.1 billion seed packets to the nation's farmers as a means of strengthening regional crop diversity and encouraging local adaptation. Seed saving was part of who we were as a culture and nation.

How is it, then, that in such a short time we have been convinced that we must buy "new" seeds every year? And how have we come to believe that seed saving should be left in the hands of the "professionals"? Unpacking this radical (and disturbing) shift in our social, political, and spiritual relationship with seeds would require more space than we have here. But let's just say that this severing of our connection to seeds appears to be calculated, thorough, and nearly complete. Thankfully, people like Cindy are emerging to restore this life-giving connection, both in our hearts and in our daily lives.

We have close personal ties with the Seed Library movement. Rebecca Newburn, founder of the pioneering Richmond Grows Seed Lending Library in Richmond, California and a prominent voice in this book, attended our first Seed School course back in 2010. We had the opportunity to launch Arizona's first Seed Library in 2012 while with Native Seeds/SEARCH in Tucson. We helped and supported the Pima County Seed Library in the Pima County Library system as they rolled out 8 Seed Library branches with inter-library loans! Their head "seedkeeper," Justine Hernandez was another Seed School graduate.

We have watched with joy and excitement as interest in this community-supported "seed solution model" has exploded nationwide. At the time of this writing, Rebecca counts more than 340 seed libraries operating in the US, with dozens more on the way. *Seed Libraries* offers an enlightening and highly instructive account of the evolution of this burgeoning movement and its passionate followers. The book serves as an excellent primer with step-by-step instructions for fully engaging your community in this start-up venture. Chapters include *Packaging, Signups and Other Details, Attracting Patrons* and an important *Resources* section. Cindy did such a good job unveiling the evolution of Seed Libraries and the many elements involved in creating one, that most of the barriers to entry have been removed. However she still cautions us about some of the stumbling blocks to a successful collaboration such as volunteer burnout.

As we continue to watch the disappearance of diversity, consolidation of the seed industry, and unbridled, exponential growth of genetic manipulation, it is apparent that a seed library grass-roots movement is among one of the best ways to create a parallel reality, one rooted in the ancient tradition of seed saving. Often we are asked how we stay optimistic as we witness the ongoing take-over of our seeds here and abroad. If it weren't for seed libraries, small burgeoning seed companies, seed exchanges, and those reconnecting with traditional seed breeding we wouldn't. But everyday a new seed library opens, books like this one are published and new seed organizations are established such as the Rocky Mountain Seed Alliance, which we were fortunate to cofound with an old friend. It is the energy and momentum of the seed saving movement, so powerfully depicted in this book, and the incredibly captivating stories told in these pages that raise our spirits and give us hope.

Seed Libraries is a new kind of treatise on who we are as farmers, gardeners, citizens, and human beings. It brings into focus an essential but too often neglected truth: that our freedom to steward our own seeds is an inalienable right as fundamental as life itself. To bureaucrats and corporatists who would challenge that right, we say: *beware*. We Americans have been known to get a little touchy when our liberties are threatened. How this story unfolds is still up for grabs as the traditional seed saving and Seed Library movement explodes. With this heartfelt and timely book, Cindy Conner has done a masterful job of capturing the life-affirming spirit and promise of our own powerful chapter in this developing tale of seeds and the people who love them.

Introduction

IN THE LAST CHAPTER of *Grow a Sustainable Diet*, I wrote about the seed library our daughter, Betsy Trice, had started at the community college where she teaches sustainable agriculture. At the time, I didn't know that I would soon be working on a whole book about seed libraries. At the urging of Ingrid Witvoet, my editor at New Society Publishers, that's just what I did. Thanks, Ingrid and the staff at New Society! It has been quite a journey.

Saving seeds was once a necessity of growing food. When seeds became commercially available, gardeners and farmers took advantage of that convenience, just as families today take advantage of buying prepared food, rather than cooking at home. There is a danger when we relinquish knowledge and skills in exchange for convenience. We are put at the mercy of whoever is managing that convenience. In the case of seeds, it is large corporations (chemical companies, actually) that are in control — through patenting and genetic manipulation. We don't have to stand idly by and let this happen. We can go back to saving seeds ourselves. However, once knowledge and skills have been forgotten, it is not so easy to revive them. Seed share programs are popping up everywhere in the form of seed libraries, or more informal seed swaps.

The people running these programs will need some direction, and that's where this book comes in. I have been a seed saver for many years and can foresee the challenges that someone starting a seed library might face.

When I asked my friends who are involved with seed companies what they thought of seed libraries, their first response was usually the question "Do people really bring seeds back?" When I talked with seed librarians in their first year with their seed library, they often had the same question. A new seed library needs to find a balance within the community. Not everyone will bring seeds back, but some will, and the library can build on that. Hopefully, a seed library will develop into more than just trading seeds. Whether we know it or not, our whole culture is wrapped up in seeds and they can be shared and celebrated in many ways. You will find ideas about how to do that in this book.

Rebecca Newburn developed the Richmond Grows Seed Lending Library in Richmond, California, in a way that could be replicated. You will be reading about how to tap into her resources. The Sister Seed Library list that she maintains was a big help to me in my research. Thank you, Rebecca! And thank you to all the seed librarians who took the time to verify what I wrote about their projects and offer me additional information. I was able to visit some of the libraries on the list, and you will find photos of them in this book. I appreciate the time and the experiences those seed librarians shared with me. What I found was that most seed libraries are in their early stages and are discovering that they need to make changes as they move forward. It is important to stay flexible.

Besides seed libraries, other seed sharing projects are emerging. You can read about some in Chapter 10. It is as if the world is awakening to the importance of seeds. With that awakening comes the need to learn everything there is to know about seeds, particularly how to save them properly. However, saving them is only part of the adventure. Distributing them to people who will grow them out and save them again is the other part. Although I don't go into the details of how to save seeds, this book will help you understand how to establish your seed project and keep it going. You may even decide to go to *seed school* to learn more. Bill McDorman and Belle Starr operate a seed school, and it is fitting that they have agreed to write the foreword for this book. Thank you Bill and Belle! (Bill co-authored the article that my daughter read that prompted her to start the seed library at her community college. Once the seed library was launched, I wrote a blog post about seed libraries that caught the attention of Ingrid. It's funny how things work out.)

We need to be mindful about what we do and how it affects those around us — and the generations to come. Everything is connected. In Chapter 1 you will see a photo of old hands pouring seeds into young hands. It is actually my old hands pouring seeds into my five-year-old grandson's young hands. When we choose to work with seeds, we direct what is going to go forth from there. It is both a daunting and an exciting venture.

Some people think that the systems in our society are breaking down. I guess that's one way of looking at it. In reality, systems need to change to stay relevant to the conditions at hand. It takes a special person to venture out of the norm and be the catalyst of a change that needs to happen. Thank you to all the pioneering souls who have already ventured into the realm of putting seeds back in the hands of the people. Also, thank you to those just beginning the adventure — the readers of this book. Together we will make a difference.

CHAPTER 1

A Growing Movement

Plants have ingenious ways to produce and disperse their seeds without any help from mankind. They are capable of adjusting to their surrounding conditions and climate to ensure the continuation of their species. Once we humans came into the picture and started domesticating plants, expressing control over agriculture, we began to save seeds. In the process, crops were gradually changed in ways we thought would benefit us; we started selecting for size, color, timing, hardiness, etc. Farmers would save their own seed and trade with their neighbors. Beginning about 1790, the Shakers began selling vegetable seeds. They were among the first to do so, and were certainly the first to sell them in paper envelopes. Inevitably, more seed companies began operations and people willingly stopped saving seeds in exchange for the convenience of buying them. At first, the seed companies they bought from were small companies selling seed regionally. Eventually, the small seed companies were bought out by larger corporations that marketed seed nationally.

Handing the safekeeping of our seeds over to corporations is sort of like what's happened with soup. Soup is such an easy and natural thing to make from leftover ingredients and bones. Once canned soup came on the market, many leftovers and bones went to the compost pile or worse yet, to the trash, without having another go-around on the dinner table. Soup eaters were put at the mercy of whatever the soup companies wanted to include, which might be an overload of salt or other unnecessary and harmful ingredients. Not only did we give up control over a basic food in our diets, we lost the goodness extracted from all those bones and leftovers. But perhaps even more importantly, we lost the knowledge of how to make soup. In the growing of plants, we have the ability to reap the benefits of all they have to offer, but when we stop short at saving seeds, we are at the mercy of whatever is offered for sale. Sometimes our favorite varieties are discontinued — gone forever unless we save the seed ourselves.

The large seed companies are looking to sell seeds to as wide a demographic as possible, so varieties that do well only in limited areas are discontinued. That's the way big business does things. The corporations that now own the majority of seeds did not start out as seed companies, but as chemical companies. They market seeds that go well with their chemicals. In 2009, ten companies accounted for 73 percent of the global commercial seed market. Three of those companies alone controlled 53 percent of the global market. Monsanto, the world's largest seed company and the fourth largest pesticide company, controls 27 percent of the total global commercial seed market.[1]

Large corporations have been buying seed companies since the 1970s.[2] Furthermore, they have modified the genes of some crops by inserting genes that are not even the same species. The possibility exists that something is genetically added to a variety that produces allergic reactions in some people, among other horrors. Those crops are designated as GMO (genetically modified organisms) or GE (genetically engineered). Making matters worse, presently GMO foods are not required to be labeled as such. Monsanto is the most well known for GMOs, particularly with their crops engineered to withstand glyphosate, the active ingredient in Monsanto's Roundup. Roundup Ready crops can be sprayed with the herbicide to kill weeds around them, but the crop won't be damaged. It will be covered with herbicide, but it will survive. Weeds adapt to persist in the environment and are adapting to

Roundup in the form of superweeds that are not affected by the herbicide, requiring other measures to be taken.

Monsanto crops also include Bt corn and cotton. Bt (*Bacillus thuringiensis*) is a bacterium that produces crystal proteins (cry proteins), which are toxic to many species of insects. Bt has long been an option for organic growers to apply selectively to their crops for insect control. However, with these GMO crops, Bt is in every cell of the plant, creating an overload to the ecosystem and to our bodies. This insecticide cannot be removed by washing or peeling when it is part of the Bt crop and will add to the accumulation of insecticides in our bodies when we eat the food. Insects that Bt was intended to control have adapted to it, just as the weeds have adapted to Roundup. In addition, where Bt did its job of controlling certain insects, secondary problems arose, with new insects moving in.[3]

GMO corn does not improve nutrition. DeDell Seeds, located in Canada and one of the few "GMO-Free" seed corn companies, offers a study[4] that shows the nutritional differences between GMO corn and non-GMO corn. According to the study, non-GMO corn has 437 times

The Safe Seed Pledge:

"Agriculture and seeds provide the basis upon which our lives depend. We must protect this foundation as a safe and genetically stable source for future generations. For the benefit of all farmers, gardeners and consumers who want an alternative, we pledge that we do not knowingly buy, sell or trade genetically engineered seeds or plants. The mechanical transfer of genetic material outside of natural reproductive methods and between genera, families or kingdoms, poses great biological risks as well as economic, political, and cultural threats. We feel that genetically engineered varieties have been insufficiently tested prior to public release. More research and testing is necessary to further assess the potential risks of genetically engineered seeds. Further, we wish to support agricultural progress that leads to healthier soils, genetically diverse agricultural ecosystems and ultimately healthy people and communities."

the calcium as GMO corn! GMO cotton has also not lived up to the promises made to growers by Monsanto, particularly in India. (More about that in Chapter 10.)

In 2000, I had the pleasure and privilege to hear Dr. Vandana Shiva speak at the PASA (Pennsylvania Association for Sustainable Agriculture) Conference. She is a scientist and environmentalist from India who has dedicated her life to fighting patented and GMO seeds and promoting organic farming. Dr. Shiva founded the organization Navdanya[5] to further her work. Navdanya promotes non-violent farming, which protects biodiversity, the Earth, and small farmers. This organization works toward conserving seeds with climate-resilient properties, including 4,000 rice varieties and 195 wheat varieties, plus other cereals, pulses, millets, pseudocereals, oilseeds, and medicinal plants. The existence of Navdanya and many other enlightened seed companies proves you don't have to settle for seeds brought to you by large corporations. Some seed companies have even signed the Safe Seed Pledge to "*...not knowingly buy or sell genetically engineered seeds or plants.*" Look for that information in their catalogs or on their websites.

Hybrid vs. Open Pollinated Seeds

You will find some seeds designated as hybrids. Hybrids are a cross between two parents of different varieties of the same species to develop a variety with desirable traits that the parents don't have. You can save the seeds from a hybrid plant, but when you grow them out, you won't necessarily get plants with the desirable characteristics. For those special characteristics, you need to go back to the seed company each year for hybrid seeds. For a seed company that is in business to make a profit, that's good business. Varieties with the designation *F1* near the name are hybrids.

I grow open pollinated plants. The seeds saved from those varieties will breed true, as long as I have taken precautions to prevent them from cross pollinating with something else. That's the tricky part — knowing what will cross with what and keeping them separate. Depending on the method of pollination (self, insect, or wind), plants of the same species could cross with each other. It is best to begin your seed saving with self-pollinators. Chapter 7 contains more information, including the skill level involved with each crop. You will find books on seed saving that will help you with those details (and more) in the "Resources" section of this book.

Heirloom varieties, sometimes referred to as heritage varieties, are open pollinated. Some small seed companies specialize in open pollinated seeds. They can stay in business because so many people don't want to be bothered to save seeds; however, once you buy them you can grow them out and save them yourself. As I mentioned before, seeds evolve with their conditions. Buying seeds that have been grown far away means your seeds have to start new each year, with the seeds having to adjust to your climate. If you or someone relatively near you save seeds each year to plant back, those seeds are already conditioned to do better in your garden.

Sometimes open pollinated varieties are intentionally crossed to make new varieties. The first generation would be an F1 hybrid, but if it is selectively grown out for seven years (or so), it could become a stable open pollinated variety. An open pollinated variety can itself be worked with, selecting for specific traits that are already present, although not yet uniform, until the desired traits are stable. You can even do this in your own garden! It basically consists of just pulling up any plants that don't have the characteristic you want. You will find books on plant breeding in the "Resources" section to help you through the details.

Some seed companies spend much time and money breeding varieties with special traits and, in the spirit of making the most of the research that has gone into developing new varieties, some varieties have *PVP* designations. PVP stands for *plant variety protected.* The Plant Variety Protection Act went into effect in December 1970. "Its purpose is to encourage the development of novel varieties of sexually reproduced plants by providing their owners with exclusive marketing rights of them in the United States."[6] PVP varieties are protected for 20 years (originally it was 18).

I first became aware of the PVP designation when I was a market grower. For a couple years I had grown Romulus, a variety of romaine lettuce with an open head. I liked the open head because it was easy to clean for my restaurant customers. Although there appears to be many open-headed romaine varieties on the market today, at that time it was the only one I could find. When Romulus wasn't offered in the seed catalog one year, for reasons I can't remember, I decided to grow it out from seeds left from the previous year. It was my first experience growing lettuce to seed, and I was pleased with my harvest of four ounces of seed. When the seed catalogs arrived that next winter, I saw Romulus listed by another seed company with the letters PVP beside it. When I discovered what that meant I was worried about what it meant for me — if I wanted to continue as a law-abiding citizen. I found out that I could save my own seeds and grow them out; I just couldn't sell the seeds or give them away. I was growing it in the 1990s. The PVP protection expired for Romulus in 2010, making it fair game for any seed company to propagate and sell. It was a great romaine variety, but you will have a hard time finding it today. What you *will* find are new varieties that the seed companies have control over. A seed company that has developed a PVP variety puts resources into marketing it and keeping the variety strong while they have an exclusive right to it. By the time the PVP expires, they have developed new varieties; these are given the same special attention, and nothing is done to further the development of the expired variety. That's business for you.

You will likely come across these PVP varieties when you see varieties with novel claims. The seed companies are required to label them as PVP, so you can be on the lookout for them (although when I first bought Romulus I didn't see a PVP label). You can find which varieties hold PVP designations now and in the past by accessing the Certificate Status Database[7] maintained by USDA's Plant Variety Protection Office.

Some seed varieties are covered by a *utility patent*. You are *not* permitted to save the seeds of those varieties for replanting yourself.[8] Utility patents last for 20 years from the date of filing with the United States Patent and Trademark Office.[9] If a plant is a PVP variety or is covered by a utility patent, that information will be mentioned in the seed catalog that carries it. Read variety descriptions carefully when ordering from a seed company.

Grassroots Seed Saving and Sharing

We can opt out of participating in what big business has to offer. We can save our own seeds and share them with others. Seed Savers Exchange in Decorah, Iowa, began with that very idea. Diane and Kent Whealy wanted to keep the seeds of Diane's grandparents in production and share them with others. From that humble beginning grew the widespread network of seed savers that exists today. In Canada, Seeds of Diversity, formerly known as the Heritage Seed Program, is a volunteer organization that conserves the biodiversity and traditional knowledge of food crops and garden plants. Through the *Seed Savers Exchange*

CREDIT: JAROD CONNER

Yearbook and Seeds of Diversity's *Member Seed Directory,* members of these organizations can obtain seeds from each other, making available thousands of varieties that are no longer found elsewhere.

In the past few years, people have realized the importance of preserving seeds and have begun to organize seed libraries. Seed libraries are a fast-growing movement to foster seed saving, allowing gardeners to be involved in the most basic part of the production of their food. Seed *banks* are repositories that hold seeds for the future. Seed *libraries,* on the other hand, are dedicated to getting seeds to as many gardeners as possible to be grown out each year — allowing the varieties to be preserved, while at the same time adapting as needed to the local climate and conditions.

The libraries house the seed, provide resources, and offer classes teaching patrons to save seeds. The gardeners save seeds and give some back to the seed library. Getting seeds back that have been grown properly — without cross pollinating with something else — is a concern for a seed library. If cross pollination occurs, however, it can open the way for a new regional variety to develop. A market gardener friend of mine who lived in Montpelier, Virginia, saved cucumber seeds one year; when he grew them out the next year, the cucumbers happened to be white, rather than green. They still tasted like cucumbers, so he labeled them Montpelier Whites and sold every one. Stay open to new possibilities.

We usually think of a library as a brick-and-mortar building where we can borrow books or use the Internet. It keeps resources safe, but available when we need them. We borrow books from a library, but we have to bring them back in a timely manner. When libraries offer seeds, how does that work?

Seeds "borrowed" from seed libraries are not expected to be returned anytime soon. The idea is to plant the seeds and grow them to maturity, hopefully saving and bringing back as many or more than you "borrowed." Seeds are returned at the end of the season. But things don't always work out as planned. The weather doesn't cooperate, insects or other predators take out the plants, your seed saving skills are not what you thought they were, and the list goes on. The founders of seed libraries anticipate that not everyone will be able to bring seeds back. They know that others will be able to bring back more than they borrowed. There are many ways that individual libraries are set up, and I will be telling you about them. If you have never saved seeds before, don't let that stop you from participating. Everyone has to start somewhere. Sometimes seed libraries are located within traditional book libraries, but that's not the only place you'll find them.

The first seed library to form was BASIL — Bay Area Seed Interchange Library — which began in 2000 at the Berkeley Ecology Center in Berkeley, California. BASIL is sponsored by the Ecology Center and run by volunteers. Sascha DuBrul came up with the idea of a seed library when he wanted to find a home for seeds that were left when the University of California, Berkeley, closed its campus farm. (The property was to be used for cooperative research with Novartis, a Swiss biotechnology corporation.) Christopher Shein was the farm manager for a one-acre plot at the farm that had been used by an organic farming research class to grow food and seeds to share and to preserve heirlooms. Shein and Terri Compost worked with DuBrul to establish the seed library. DuBrul eventually migrated to upstate New York where he spread his enthusiasm for seed sharing to Ken Greene, a librarian at the Gardiner Public Library. Greene started a seed library there in 2004. In 2008, it became the Hudson Valley Seed Library and moved to Accord, New York, operating on a different model altogether under the direction of Greene and his partner Doug Muller. You will be hearing more about the Hudson Valley Seed Library in Chapter 10.

Projects evolve to suit their communities. If something doesn't seem to be working as envisioned, it can be changed to make it a more viable activity. In 2007, Caitlin Moore started the Olympia Seed Exchange in Olympia, Washington. Within a year, she was joined by Claire Ethier, and later by Casey Fabing. Originally begun as a website, it has changed and evolved over the years and now operates by hosting monthly seed exchanges at different locations around town, teaching classes on seed saving, and participating in local and regional farming and seed-related events.

SPROUT Seed Library was established about 2006 in West Marin, California. SPROUT, an acronym for Seed and Plant Resources

OUTreach, provides seeds and plants to gardeners who, in turn, grow out select plants for seed and return some to the library for others to borrow. The SPROUT library has a portable component to it. SPROUT offers classes and seed saving resources.

About 2008 the Portland Seed Library was organized, hosted by the Northeast Portland Tool Library, which has its home in the Redeemer Lutheran Church in Portland, Oregon. By the end of 2010, the Seed Library of Los Angeles (SLOLA) was formed using The Learning Garden at Venice High School as its base.

Influenced by a permaculture design course she took (taught by Christopher Shein), Rebecca Newburn co-founded the Richmond Grows Seed Lending Library (Richmond Grows) with Catalin Kaser in 2010. Newburn wanted the seeds to be more accessible to people, and came up with the idea of putting them in an actual public library, independently of what Ken Greene had done in New York. Her goal was to make a replicable model for others to follow. The home for Richmond Grows is the Richmond Public Library in Richmond, California. You

will be hearing more about Richmond Grows in Chapter 4. Anyone researching how to start a new seed library probably already knows about Richmond Grows through its website www.richmondgrowsseeds.org. Although some information will continue to be there, the new go-to place for seed library information is www.seedlibraries.net, also managed by Rebecca Newburn. Newburn is a middle school science teacher, and her enthusiasm for seed saving has moved into her classroom curriculum. Since 2010, seed libraries have sprung up at a steady rate, with the pace quickening in 2012 and growing faster each year. You can find a list of known seed libraries, referred to as Sister Seed Libraries, at www.seedlibraries.net.

In the News

Seed libraries are newsworthy events. In March 2013, seed libraries were the subject of a segment on the *NBC Nightly News.* That was the same month the seed library opened at J. Sargeant Reynolds Community College in Goochland, Virginia. My daughter, Betsy Trice, had been influenced by an article in *Acres USA* the previous year and, with cooperation from the college library and donations of seed from seed companies, made it happen at the college. That article, "Sowing Revolution: Seed Libraries Offer Hope for Freedom of Food,"[10] by Bill McDorman and Stephen Thomas was the result of a gathering of seed savers at the National Heirloom Exposition in September 2011. Word gets around. Ordinary people catch an idea from somewhere, whether from TV, in print, or from a conversation with someone, that sparks something within them to act. Without too much trouble, you can find Internet postings and news articles about start-ups of new seed ventures. People are coming together to help each other save and share seeds, taking control of the most basic part of their food supply. The more seeds are shared, the better they are preserved in the public domain.

Find Seeds Native to You

Native Seeds/SEARCH is a nonprofit conservation organization based in Tucson, Arizona, that has been collecting and dispensing seeds of crops indigenous to the southwestern United States and northern

Mexico since 1983. Native Seeds/SEARCH also offers seeds of crop varieties that are not traditional to the Southwest, but that can contribute to regional food security. It has become a major regional seed bank and a leader in the heirloom seed movement.

What is indigenous to your area? It may be that some of the crops and the history that you want to see preserved originated with immigrants who arrived with seeds in their pockets, coat hems, and hatbands from their homelands. Sometimes the only connection immigrants have with their heritage once they have been transported to a new country is through food, which ultimately means through the seeds that grow it. The stories connected to the seeds are often just as important as the seeds themselves. There is always a story that serves to connect us to what has gone before. In this new wave to save seeds, we should be sure to also save the stories.

CHAPTER 2

Why Save Seeds?

Protection from corporate domination is the driving force of most seed saving initiatives today. In Chapter 1, you learned how much control a few companies have over our seed supply. From the seeds comes our food. *Whoever owns the seeds controls the food supply.* If you remember one thing, remember that. Rather than falling into a black hole of depression over the thought of being controlled by Monsanto or companies like it, we can empower ourselves and save seeds of what we grow and share them with others, keeping them in the public domain and available to everyone. We can opt out of corporate control and build a new system. Say "No thanks, I know there's a better way." There are many more reasons for saving seeds besides avoiding corporate domination, although many of these reasons stem from the corporate takeover of seeds.

> Whoever Owns the Seeds Controls the Food Supply

Preserve Genetic Diversity

A broad genetic base is necessary to keep plant populations strong. When problems occur or growing conditions change, it is good to have a wide variety of genes in the mix to come to the rescue. It's dangerous to depend on only a few varieties of a crop. Each one has its own strengths and weaknesses. A variety known for its huge yields won't produce them every single year. Growing different varieties evens out the yield, with some doing best in one year and some doing best in another.

History has shown us what happens when we ignore genetic diversity in our plantings. The most well-known example is the Great Irish Famine that occurred when blight attacked the potatoes in 1845, continuing to affect the harvest into 1849. Genetic diversity was certainly lacking in Irish potatoes, with the harvest coming from mainly one variety, making it easy for the fungus to spread. Problems in Ireland began long before the potatoes died. The land system that had developed as a result of wars and politics left Irish laborers and their families in poverty, often existing solely on potatoes. Other food was grown in Ireland and even exported during this time; politics, such as they were, preventing it from being shared with the hungry. Some of you reading this are probably descendants of the 1.5 million Irish who left their homeland to escape the famine. The blight affected potatoes in other European countries, but it was the Irish peasants who were the most dependent on potatoes for their nourishment. You can find out more about the Great Irish Famine in the book *Black Potatoes*[1] by Susan Campbell Bartoletti.

In 1970, the US corn crop was hit hard by blight, reducing the total harvest across the country by 15 percent. In the Southern states, the harvest was reduced by 50 percent. This was caused by a fungus that affected hybrid corn. At the time, hybrid corn seed contained *T-cytoplasm,* which carried a gene that opened the door to the fungus. "T-cytoplasm was a man-made change in corn plants used to foster the quick and profitable production of high-yielding, hybrid corn seed."[2] Originating in Florida, in just four months Southern corn leaf blight spread west to Kansas and Oklahoma and north to Minnesota and Wisconsin; it later entered Canada. In a more holistic approach to corn growing, varieties chosen to plant in these diverse areas would be different. In a perfect world, the varieties planted would be open pollinated varieties unique to each area. The plant breeders did not know about the potential consequences at the time, but we know now and need to prevent things like that from happening again. In *The Omnivore's Dilemma,*[3] Michael Pollan writes about our dependence on corn in our diet, whether eaten directly or indirectly. Besides genetics, we need to have diversity in so much more of our lives. It is not healthy to limit our diets to only a few crops.

Much has been said about the loss of seed varieties, and the resulting loss of genetic diversity. Sadly, 57 percent of the nearly 5,000 non-hybrid vegetables varieties offered by mail-order seed companies in 1984 had been dropped by 2004. Consolidation within the mail-order seed industry and a move to more profitable hybrid varieties are among the reasons for that loss. But an astonishing 2,559 new varieties were introduced during the period 1998–2006 alone. You will find these figures and more in the 6th edition of the *Garden Seed Inventory*[4] published by Seed Savers Exchange (SSE). Compiling that book is a huge effort. Thanks to SSE, we have a snapshot of the open pollinated varieties offered commercially since 1981. Records beginning in 1987 show more varieties are added each year than are dropped. That's promising! Small specialty companies and individuals are keeping the varieties alive and developing new ones. Also contributing to the increase are the varieties from other countries that have been made available in the US and Canada. The genetic diversity, so necessary to maintain, lies in the hands of individuals and small seed companies.

Plants are such wonderful things to work with. Keeping them in production is the best way to preserve varieties and to have genetic material available to work with during changing times. The expression "use it or lose it" definitely applies to seeds. The varieties recorded in the *Garden Seed Inventory* are the ones available commercially. But there are so many more seed varieties being grown in the world, and Seed Savers Exchange and Seeds of Diversity Canada are places to look to for them. If you want a peek into what is possible, check out their member directories. I visited the Summers County Public Library Seed Lending Library in Hinton, West Virginia, when it was just getting established, and the librarian there expressed a desire to acquire varieties of seed from area residents whose families may have been saving them for generations. Seed libraries are excellent places to accept and disperse these heirloom seeds.

Preserve Flavor and Nutrition

Genetic diversity naturally leads to diversity of so many other things. Most people are familiar with tomatoes, so I'll use that as my example. Tomatoes grown for sale through grocery stores need to be the all the same size to fit nicely in boxes and to withstand the rigors of going through the picking-washing-packing-shipping-holding process. Flavor and nutrition are not high on the list of criteria for choosing grocery-store varieties. But flavor and nutrition are very high on the list for many home gardeners. For maximum flavor and nutrition, it is important to leave tomatoes on the vine as long as possible to ripen. As soon as you pick anything, it begins to lose nutrients. If you pluck

Tomatoes ripening on the vine. CREDIT: BETSY TRICE

tomatoes off before they are ripe, they never get a chance to gain their full nutrition. By growing them yourself, you can ensure the best flavor and nutrition. By saving seeds, you can make sure to have a particular variety in the future. Tomatoes selected for flavor and nutrition may not be the "prettiest," and they just might be easily bruised if you keep them on the vine until fully ripe. That's okay if you are bringing them into your kitchen to use, but not okay in the industrial food chain. Tomatoes destined for supermarket shelves are picked while they are still green so they can withstand the trip.

Vegetables come in many different colors and shapes that you never see in the grocery store. You can try them all in your garden and save the seeds of the ones you like best to make sure you have them to grow again. Have you ever seen purple carrots and red okra in the grocery store? The different colors offer a slightly different mix of nutrients. I love it when my food has extra visual interest, but for me it ultimately all comes down to taste. I grow a variety of large cherry tomato that has a real tomato flavor that I like. There are some sweet-tasting cherry tomatoes in the marketplace that I don't care for, but some people love them. If seed companies decided to discontinue all the cherry tomatoes except for the sweet ones, I would be out of luck if I didn't save my own seeds. An interesting culinary challenge would be to find as many different varieties of a crop as you can to use in your cooking and discover the nuances of each. When it comes to saving seed, however, you would need to limit the varieties grown or take precautions so the varieties don't cross pollinate.

In recent years, I heard a news report on TV that said our food had less nutrition than it did 20 years ago. I took notice because I remembered reading in a book[5] published in 1983 that the nutritional quality of our food was declining. While doing research for my book *Grow a Sustainable Diet,*[6] I found further evidence of declining nutritional value of the food that is available through commercial sources. This decline has been going on for some time now, and nothing has been done to stop it. One reason this is happening is the declining quality of the soil that crops are grown in. Building your soil is a whole other discussion, so I will limit myself to seeds here. Another reason for declining nutrition is that nutrition is not necessarily one of the characteristics selected for with seeds that go into commercial production. Your choice of seeds makes a difference.

Preserve Unique Varieties

In your garden, you might want varieties that will ripen over a longer period of time for an extended harvest. On the other hand, if you are preserving in quantity, you might choose varieties that ripen all at once. "Grows well in dry times" or "grows well in wet times" might be considerations. When I was choosing a pole lima bean to grow, I chose one that was an heirloom variety and was listed as the hardiest of the choices. The description also said the pods of this variety might pop open when they are dry, dispensing the seeds. Since I let the pods dry on the vine, I watch for this. Each variety has its own special characteristics.

Sometimes seed companies can't find a grower every year for certain varieties or the seed produced was inferior for some reason. The next year, that variety might not be listed. Taking a look at a seed catalog now, I see that 5 varieties out of 11 pole limas listed are unavailable for the coming year. I tried St. Valery carrots for the first time two years ago and liked them. I didn't save the seeds and was disappointed they didn't make it in the catalog last year. I see that St. Valery is offered this year. Don't take it for granted that the varieties you want will always be there.

Years ago, my friend Kevin wanted an open pollinated variety of pepper that ripened to orange. Not being able to find one, he bought some hybrid pepper plants and saved the seeds. He had to carefully select each year for the characteristics he was looking for, but eventually he stabilized the seed and gave me some. That is the orange pepper that I grow. One winter, I received a call from the woman who wrote a weekly gardening column for the newspaper. She wanted to know the varieties of different vegetables that I grew. All I could think of was that she was going to publish this and people were going to take that list as gospel, even though I might choose to grow something else in later years. (Sure enough, people stopped me in the grocery store more than once to tell me they were growing the varieties I named.) When she got to peppers, I told her my current favorite was Kevin's orange pepper, but that she wouldn't find it in the seed catalogs. I related the story of how I came to grow it, and she asked how others could get it. I said to go the farmers market in August, buy orange peppers from Kevin, and save the seeds. Well, she loved the story and printed it. She had written about Southern Exposure Seed Exchange (SESE) in the same article. About the time the article was published, I was at a conference, and my friend Ira from SESE said she would like to get some seeds for the orange pepper. I introduced her to Kevin at the conference. Soon after, he gave me seeds that I passed on to Ira. SESE grew them out and liked them well enough to add Kevin's Early Orange pepper to their catalog.

That's how things happen in my world. It doesn't take large corporations with huge research facilities and budgets to make the world go around. People following their hearts and passion can make a world of difference.

Preserve Cultural Heritage

Seeds and plants connect us to our heritage and place. Some folks begin to save seeds when they are passed on to them by family, particularly when that family member has kept them alive over generations. Saving seeds from one year to the next was once a way of life. That meant that gardens were planned with seed saving in mind. Growing to save seeds was just as important as growing for food for the table.

Have you ever taken a bite of something and been transported back to a time and place that held fond memories? Maybe it's a memory of visits with your grandparents when you were a young child. Food can do that to us. The immigrants that come to this country bring their culinary tastes with them. Living in a strange land, they find comfort in foods from home. That is evident from the ethnic grocery stores and restaurants that proliferate in neighborhoods with a large immigrant population. I imagine the seed libraries that are established in those neighborhoods will have seeds of the foods that the residents know from their homeland.

You don't have to be from another place to enjoy the cultural heritage of someone else. The more we get to know one another, the better we understand each other. Food is a good starting place. Besides the restaurants and grocery stores, some communities have annual food festivals that are open to all to enjoy. There were no seed savers in my family, so it will start with me. It has already extended to the next generation, with my daughter giving me seeds of Turkey Craw beans she had grown, a variety that I had never tried. They did so well in my garden last year that they will be in there again this year. Best of all, I think of her when I work with the seeds, just as I think of Kevin when I grow orange peppers. My work is mostly with vegetables, but flowers and herbs can affect us the same way. With seed libraries, the seed saver's name is on the seeds they bring back to share. You can think of the seed savers when you grow them out.

Develop Strains Unique to Your Microclimate

Getting just what you want to survive in the microclimate of your garden is what you are after. If you save seeds from the plants that do the best in your garden, you will be developing a strain of that variety that is particularly suited to your conditions. When immigrants brought seeds from their homeland to plant here, most likely the growing conditions were different from where they came from. The seeds and plants that survived would have gradually become acclimated to the new land. I've heard of people moving from one part of the country to another, taking saved seed with them. Over time, the plants became acclimated to the new area and had slightly different characteristics than where they grew before.

I gave up starting seeds indoors under grow lights years ago in favor of starting everything in coldframes. The only challenge I had was with peppers. Pepper seeds like the soil to be a little warmer than what the coldframe provides when I'm planting them. I figure that, even if I have poor germination, the plants that grow will be the ones better adjusted to those conditions and worth saving seeds from to breed my own strain of coldframe-tolerant peppers. My favorite pepper for starting in the cold frame is Ruffled Hungarian. In 2008, I happened to buy a few plants at a health food store that had been grown by a local nonprofit. I had never seen this variety before, so I thought I would give it a try.

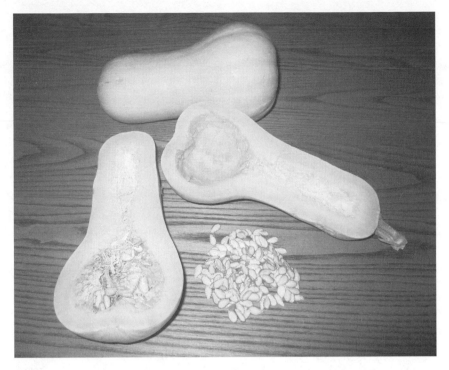

Butternut squash with seeds. This photo was taken in June. The winter squash was grown the previous season and stored in the house through the winter.

Ruffled Hungarian is a thick-walled, sweet pepper that ripens to red. I saved the seed and discovered that it performed better in the coldframe than the other pepper varieties I was working with.

Another favorite of mine to save seeds from is butternut squash. I store my winter squash in the house and use it as I need it over the course of the winter. I save seeds from the squash that have held up the best in storage all winter.

Attract Beneficial Insects

The guideline I always mention for garden management is "feed the soil and build the ecosystem." Seed saving is part of that, bringing the system full circle. On the way to producing seeds, the plants produce flowers which attract pollinators and other insects that will keep the harmful garden insects under control.

As a beginning gardener, I learned to harvest plants, such as basil, before they flowered for the best culinary flavor. It was only later that I learned about all the beneficial insects I could attract to my garden if I let the basil plants flower. Leave them just a little longer, and those flowers turn to seeds. Now I harvest some of my basil regularly as leaves, and let some of it flower. Try it for yourself this year and watch what happens. You may have noticed this already with broccoli. If your broccoli plants produced flowers before you had a chance to harvest it, step back and watch the show. The best time to observe the insects is from 10am to 2pm on a sunny day.

Some crops, such as carrots, beets, parsley, celery, and cabbage, need to overwinter in the garden, or be stored appropriately and planted out in early spring, because they produce their seeds in their *second* year. You need to plan ahead for this when making your garden plan so you have space reserved for the seed crop in the spring. The best thing

Celery flowering, on its way to seed.

is that when these plants perk back up early the next year, they will produce flowers that attract beneficials, and you won't have had to do anything but let them grow! I like to grow celery to use the leaves in cooking all summer and to dry them in my solar dryers. I make sure the plant has enough foliage to go into the fall. It will die back, but when it pops back up in the spring, besides attracting the good bugs, it produces celery seed that I can save for cooking with, in addition to save for planting.

Save Money

No doubt about it, the cost of seeds and shipping them to your home continues to climb. Saving money is as a good a reason as any to save seeds. A few packets of seeds when you are starting out doesn't put too much of a dent in your pocketbook, but as you begin to grow more of your food, you will need more variety and a larger quantity of seed.

I have always been surprised at the number of market growers who don't save their own seeds. When I sold lettuce, I let it mature and cut it as whole heads of leaf lettuce. I certainly received my money's worth from each seed. The trend now is to sell lettuce mixes that are harvested as baby leaves. Even though the plants provide several cuttings, it takes quite a bit of seed to produce that. Setting aside a growing area to let some lettuce go to seed could save you money while providing habitat for beneficials.

Learn New Skills

If you have been gardening for a while, maybe it is time to expand your gardening expertise and learn new skills. When you save seeds, you have to be aware of which varieties will cross, the timing of the seed harvest, how to get the seeds from the plant to your seed-saving container, and how to store them so they will be viable for as long as possible. Learning new things keeps life interesting.

I always encourage people to plan their garden carefully in the winter and order all the seed they need then, so there are no delays when it is time to plant. We are lulled into a false sense of security when we can pick up the phone and place an order, or order online, and within a week, the seeds show up in our mailbox. However, things can happen. When I was a market grower, I was looking to expand my fall offerings to the restaurants I sold to and decided bok choy would be good, although I had never grown it before. I checked the calendar and I had just enough time to get a crop for fall sales if I planted it in the next week. For some reason, the seed companies I dealt with on the East coast were out of bok choy, which was surprising. So, I ordered the seed from Territorial Seed Company in Oregon. Remember, I live in Virginia. Territorial Seed had always been prompt, so things would work out — or so I thought. That was the year that the UPS workers went on strike. My seed order was sent out, but it then sat in a UPS warehouse somewhere between here and there. It arrived three weeks later, too late for my plans. I don't grow to sell anymore, and I don't grow bok choy, but I do grow kale, collards, and Swiss chard. I save seed from those and never have to worry about not having them when I need them. Saving seeds adds a new level of awareness, skill, and empowerment to your gardening activities.

Make Seeds a Part of Your Life

Indigenous crops and the seeds to grow them, once a part of life for all Native Americans, have been lost in many regions. The White Earth Anishinaabe Seed Library at the White Earth Indian Reservation in Callaway, Minnesota, is working to remedy that. Established in 2012, the goal of this seed library is to promote restoration of Indigenous cultural and agricultural knowledge in the region by saving seeds. This seed library program was started and stewarded by Zachary Paige, AmeriCorps VISTA volunteer for the White Earth Land Recovery Project and a graduate of the Minnesota State Community and Technical College Sustainable Food Production program. The White Earth Land Recovery Project was founded by Winona LaDuke in 1989. LaDuke is co-founder and executive director of Honor the Earth, where she actively campaigns for a land-based economy among Native American people.

The White Earth Anishinaabe Seed Library has seeds of a unique winter squash named Gete-okosomin which they are excited to be able

White Earth Anishinaabe Seed Library

White Earth Indian Reservation
Callaway, Minnesota

Began in 2012 by the White Earth Land Recovery Project (WELRP)

Mission and About: The goal of the White Earth Anishinaabe Seed Library is to promote restoration of Indigenous cultural and agricultural knowledge in the region through saving seeds. The best way to preserve traditional varieties of seeds is to grow them out, enjoy the harvest, and spread the seeds. More hands planting and caring for these seeds gives more diversity and health to the soil and our bodies.

Currently there are three library locations, and the mission is to keep adding libraries until people simply keep seeds in their homes. For many years, Winona LaDuke, and seed keepers working for WELRP, collected seeds from all over the country; the seed library program at the WELRP is currently stewarded by Zachary Paige, AmeriCorps VISTA volunteer. The seeds are given freely to dedicated seed savers who are expected to return double the amount they received and document their location, growth, and story. Education and training opportunities in seed saving are given twice a year at the White Earth Tribal and Community College. These seeds, not only of cultural interest, need to meet the climate conditions of a cold, short growing season in the region. Therefore, many of the seeds in the collection were originally grown by tribes of the region, particularly the Ojibwe and nearby Mandan and Hidatsa tribes.

Website: www.anishinaabeseedlibrary.com Here you will find information about the seeds available and links to radio broadcasts of their "Seed of the Week" radio program about seeds in the library.

This seed project is part of the Indigenous Seed Keepers Alliance (ISKA), an active, participatory, regional seed network that includes the organizations Dream of Wild Health, the Science Museum of Minnesota, Lac Courte Oreilles Ojibwa Community College, Shakopee Farm, the Intertribal Agriculture Council, and passionate, knowledgeable seed saver individuals. ☛

A USDA Tribal Colleges Research Grant was secured by the White Earth Tribal and Community College for preserving at-risk Indigenous crops through sound breeding practices. Also a Clif Bar grant through WELRP helped bring Native Seed/SEARCH's five-day seed school "train the trainers" to Shakopee, MN, in May 2014, where three new people from White Earth/Leech Lake were trained in seed saving basics.

to share. Gete-okosomin is the Ojibwe name for "Really Cool Old Squash." It is an orange, banana-type winter squash that grows over two feet long. Seeds came to the library through LaDuke, donated by folks who had been keeping this heirloom squash in production. The seeds for this variety were originally grown by members of the Miami tribe in Indiana. Many native families in White Earth have made it a part of their diet and love the rich, buttery flavor. We should think of all seeds as a gift from our ancestors, and honor them by keeping the seeds safe and growing them each year, making use of everything they have to offer.

CHAPTER 3

The Role of Public Libraries

I F WE WANT TO BE FREE OF CORPORATE CONTROL of our seeds, and thus our food supply, we need to freely trade and share our seeds with one another in public. What better way than through public libraries? They are in most every community and always looking for ways to stay relevant to their patrons. When Andrew Carnegie put his money into starting libraries, he wanted to empower people to help themselves. As a teenager, he wanted to educate himself and better his life, but couldn't afford the cost of the subscription required at the library. Once he made his fortune in steel, Carnegie began to build libraries that would be open to everyone. He paid for the buildings, and the communities filled them with books and provided the staff. I was fortunate to be able to visit two of his early libraries while researching this book. The Pittsburgh Seed and Story Library was based in the Lawrenceville branch of the Carnegie Library of Pittsburgh (Pennsylvania), with a smaller collection of seeds at the Main Library. This seed library was started in 2012 as a result of a thesis project by Amanda West (now Amanda West Montgomery), a graduate student at Chatham University. In 2014, all the seeds were moved to the Main Library.

Carnegie Library of Pittsburgh-Lawrenceville.

Carnegie wanted the resources in the libraries he built to be free to everyone. In present-day public libraries, besides books, magazines, and computer and Internet access, also free to the public are restrooms, water fountains, lighted parking, and climate-controlled meeting rooms. If you are ever out and about and need a restroom or a drink of water, you can pop into a library. When the Carnegie Library in Washington, DC, was built in 1903, it was the only public place in the city where African Americans could use the restrooms. You might not have ever thought of these things before, unless you needed to find a meeting space for a group. Meeting places are important for people to gather to exchange ideas and learn new things. Libraries also have regular hours, another important factor in making all their resources, including seeds, available to the public.

Carnegie wanted libraries to be places that fostered change. Once people have access to knowledge, they can better direct their own lives. Public libraries are where I got my start in learning about organic gardening. There was no money in our budget for books, but I read all I could find from the library. While my little ones were in story hour (another great benefit of libraries), I would browse the books in the gardening section. Our children are grown now, and we have turned one of the vacated bedrooms into a library to accommodate the books that have come our way through the years, often from used bookstores and library book sales. Still, I continue to have questions about new things and am a regular at the public library. Libraries have funds for

Carnegie Library of Pittsburgh-Main Seed Library

(formerly Pittsburgh Seed and Story Library), Pittsburgh, Pennsylvania

Began in 2012

Mission: The Carnegie Library of Pittsburgh-Main Seed Library aims to provide a public seed bank, seed saving, and gardening classes, and a place to hear and share some of Pittsburgh's vibrant gardening history.

This seed library was established as the Pittsburgh Seed and Story Library; it was the result of a thesis project by Amanda West (now, Amanda West Montgomery), a graduate student at Chatham University in Pittsburgh. West was instrumental in establishing the seed library, and the Carnegie Library of Pittsburgh maintained it. The first two years, the larger seed collection was housed at the Lawrenceville branch, with a smaller seed collection at the Main Library. In 2014, all the seeds were moved to the Main Library, which has the most traffic.

Information about this seed library is on the "Gardening Thyme" page of the Carnegie Library of Pittsburgh website (carnegielibrary.org). Gardening Thyme is a program made possible by the Mary Jane Berger Memorial Foundation in partnership with the seed library. The garden in the front yard of the Main Library is a Gardening Thyme project.

A free-standing card catalog was used to hold the seeds at the Lawrenceville branch. A smaller card catalog, originally used to take seeds to events to publicize the seed library, is used for the seeds at the Main Library. As with all seed libraries, this is a learn-as-you-go project for the librarians. After the second year of offering seeds, they marked the seed packets with the year of acquisition and stored them in the freezer for the winter.

new books and will buy according to the interests of their patrons. If you want to promote subjects to be funded, ask for books that you would like to read but don't see on the shelves.

Staff and Friends

A library already has a staff that can manage the day-to-day lending of the seeds, a big plus in establishing a seed library. Having a seed library in a book library is definitely a learning experience. The people who learn the most are the staff. Unless they were already gardeners and the idea of adding seeds was their own, librarians don't necessarily know the ins-and-outs of handling seeds. Books can sit on a shelf for very long periods of time and are no worse for wear. Seeds, however, are living things that lose their viability over time. Storage conditions can extend the life of the seeds — or cause their demise. Since libraries are generally not too hot and not too damp, conditions are right, at least for a time, for seeds. To extend seed life even longer, the seeds can be refrigerated during the off season. You'll find more information about seed storage in Chapter 5.

Libraries usually have Friends, as in Friends of the Library groups that fundraise and sponsor interesting programs. A person who volunteers as a Friend of the Library has chosen to donate their time to fostering good works through the library. If a seed library were to be started at a public library, the Friends could be an important key to finding volunteers and funding. Friends are also the people who make all those wonderful book sales happen. Book sales are a way for the library to pass on outdated or little used resources, as well as to make some money from donated books. Libraries want to stay relevant to their communities. Through book sales, space is freed up for new items (such as seeds!). The Friends of the Library may not be the driving force or provide the labor to get your seed library off the ground; however, they are good allies to have in the process. In early 2013, there were about 60 seed libraries in the nation. In early 2014, I found 163 in the US, and the number continues to grow. Whether librarians or Friends are gardeners or not, they are sure to recognize that making space for a seed library is the way of the future.

Top: *Betsy Trice at the Goochland Community Seed Lending Library.*

Center: *Location in the library at Reynolds Community College.*

Bottom left: *Guides used to locate seeds in drawers. Information given is crop, scientific name, drawer #, skill level, and crop family.*

Bottom right: *Jar with bulk seed and envelope for patron to fill.*

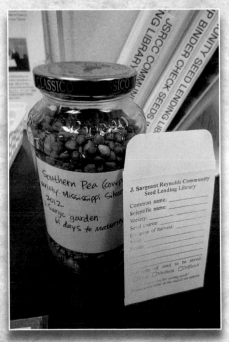

Pittsburgh Seed & Story Library
Early Purple Vienna *Spring 202*
Kohlrabi
Mustard – Root Crops

Top left: *Seed cabinet at the Carnegie Library of Pittsburgh-Lawrenceville.*

Top right: *Seed cabinet at the Carnegie Library of Pittsburgh-Main.*

Center left: *Envelope holding bulk seed. Year of acquisition was added.*

Center right: *Checkout materials are in the drawers.*

Bottom: *Notice that seeds are stored in the freezer in the off-season.*

did you know?

SEED LIBRARY SEEDS ARE BEING STORED IN OUR FREEZER UNTIL NEXT SPRING

Now it's time to save seeds and share them with other Seed Library users!

Learn how in the Seed Library zine here, in our related books, or at a seed-saving program.

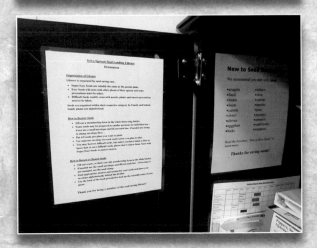

Top left: *Cardboard display rack that holds seeds at the Jackson County Farmers Market, Sylva, North Carolina.*

Top right: *The back of the display rack has seed saving and skill level information for the crops. Sylva Sprouts Seed Lending Library.*

Center: *Inside the seed cabinet at the Cooperative Extension Office.*

Bottom: *Self-guided orientation information inside the seed cabinet.*

Top: *Summers County Public Library in Hinton, West Virginia.*

Center left: *A laminated information card for each crop is kept in the drawers with the seed packets. The envelopes are pre-filled and labeled.*

Center right: *Notice to check seeds out at the front desk.*

Bottom: *Seed cabinet sits on top of a bookshelf. Vegetable seeds are on the left and flower seeds are on the right.*

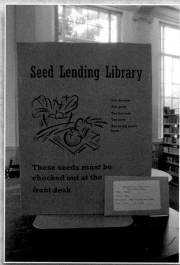

Top: *Washington County Public Library in Abingdon, Virginia.*

Center left: *Will Stein with the seed cabinet.*

Center right: *An old card catalog was painted for the seed cabinet.*

Bottom: *Seed packet used for obtaining and bringing back seeds.*

Top: *Seed sharing is a family affair at the Mother Earth News Fair.*

Center left: *Seed swap table at the Mother Earth News Fair in Puyallup, Washington.*

Center right: *Seed swap quantity suggestions.*

Bottom: *Seed swap table at the Virginia Biological Farming Conference.*

SEED SWAP
QUANTITY SUGGESTIONS

Those new to seed saving and seed swaps might want some suggestions about how much seed should be exchanged. You want to take at least the minimum amount of seed necessary to maintain the minimum genetic diversity of a variety, but at the same time you want to leave plenty of seed for other participants in the seed swap. Here are some very general guidelines for donating and taking seed:

Tomatoes, peppers, lettuce, beans (and other mainly "self-pollinated species): 25 to 30 seeds

Brassicas (cabbage family – broccoli, collard, kale, turnip, radish, etc.): 60 seeds

Alliums (onion family): 60 seeds

Cucurbits (squash family): 60 seeds

Corn: 200 seeds

Tubers, Cuttings: 2 (up to 4 if enough are available)

Plants: 2 (up to 4 if enough plants are available)

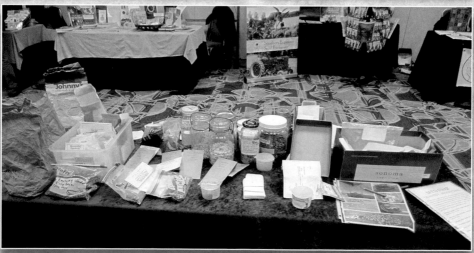

Top: *Seed cleaning screens. Cost about $190.*

Center top: *Homemade seed screens.*

Center bottom: *Interchangeable sieves found at an Indian grocery store. Cost less than $15.*

Bottom: *Collection of strainers and colanders to clean seeds. Find these in your kitchen or at yard sales.*

Top left: *Red Russian kale in a low tunnel in January.*

Top right: *Red Russian kale flowering in May. The low tunnel frame is 31" high.*

Center left: *Kale seed pods ready to harvest in June.*

Center right: *The kale flowers turned to seed pods.*

Bottom: *Kale seeds and empty pods.*

Places to Foster Creativity

If you are worried that our population has become a population of people going through life with their heads down in their digital toys, you will be happy to know that there is a movement underfoot and all around us to bring our heads up and to begin making things — whatever we can think up. Take a look at the magazines and books available today. Resources that promote things like quilting and woodworking have been around for many years. The do-it-yourself (DIY) trend is growing and is taking on a different look. The new magazines and books encourage the creative use of whatever resources you can find. One such magazine is *Make:* magazine, which had its beginnings in 2005; from it have sprung Maker Fairs and the Maker Movement, promoting transformation through innovation, culture, and education. Granted, many *Make:* projects involve technology (sometimes including a digital printer), but not all. Creativity is the overriding theme. The message: Use whatever makes your ideas work; open your mind to innovative materials and techniques to bring an idea to reality. The creativity that is encouraged crosses disciplines to allow minds, hands, and materials to work together to create something. The result is empowering to the creator. The areas carved out in libraries to encourage these activities are called *makerspaces.* I know of one public library that has added a room with a concrete floor and a drain to its renovation plans with such activities in mind.

When I think of traditional libraries, places that are neat, orderly, and quiet come to mind. When I think of myself creating, those are not be the words I would use. If I am making something from a plan or pattern, I gather the materials and put them together. You could describe that activity as relatively neat and orderly — and maybe somewhat quiet if I am alone. On the other hand, if I am creating something — making it up as I go along — I will be trying many things together, reshaping things, etc., and, in my enthusiasm, making some noise. The words neat and orderly do not describe these projects. A library staff considering the addition of makerspaces might not be completely comfortable with the changes needed to accommodate noise and an assortment of materials. But I think they would welcome the addition of seeds. Only a small space is needed and it is easily managed. (You will learn more about setting up a library space in Chapter 6.) Fortunately for the library, the mess and the noise involved in seed saving happen away from the library. The library acts as the catalyst for activity by distributing the seeds and supports it by providing books, classes and other resources. Then, at the end of the season, the library is there to receive the new seeds, setting the stage for the next season to come. Seed libraries are great makerspaces.

When beginning any project, the people involved are bound to have different ideas. I happen to be a quilter who enjoys using fabric leftover from making clothes, and the old clothes themselves in my quilts. I know other quilters who plan each quilt carefully, often following a set of directions, and buying new fabric according to their plan. We are at different ends of the spectrum when it comes to putting together a quilt. They would probably be frustrated if I were heading up a project and brought in a pile of fabric scraps and old clothes and said "Let's get started." On the other hand, I would be rather bored working on a project where all the fabric was color coordinated and bought new. There are different ways of approaching a project, even when the goal is the same — in this case, making a quilt. Woodcarvers are the same way. You can buy a piece of wood and directions for carving something, but the woodcarver I am most impressed with is the one who sees what is inside the wood and carves away the excess to expose it. It is important to know how someone's mind works before jumping into a program, whether it is your own mind or the mind of the person you are working with. If you understand that, you can better address the issues that come up when implementing a plan. The person whose passion is sparked by the very thought of a seed library needs to be the one to make it move forward, rather than someone who has been assigned the task, but who has little understanding of what is involved. However, all may have ideas that will contribute to the overall success of the library.

I first became aware that public libraries can loan more than books and magazines when my college housemate brought home works of art to hang for a month in our apartment. Now I've found that libraries loan lots of things: tools, toys, fishing poles, nature backpacks, telescopes, novelty cake pans, kitchen equipment, knitting needles, sewing machines, and musical instruments. A lending library collection of any of these items can be supported by speakers, demonstrations, classes, and books, magazines, CDs and DVDs related to them. Libraries that evolve to suit their communities will keep their patrons coming back.

Part of the Landscape

A seed library is a repository of seeds to lend, not a seed-growing operation. But, unless a library is located in a concrete jungle, there is usually some landscaping around the building. Seed-producing plants can have a place there. Given any spot of land, a garden can develop. The Carnegie Library in the heart of Pittsburgh has a garden in its front yard. The garden project and the seed library developed independently but simultaneously, which allowed a crossover of activities and of ages involved. A teen program made salad dressings and soup from the herbs from the garden, and patrons shared in the vegetable harvest. Some of the plants could have been let go to produce seeds. Having some plants for people to watch the progression from seed-to-flower-and-back-to-seeds is wonderful. Other venues housing seed libraries may be temporary or occasional, ruling out the opportunity for such a long-term activity as a garden.

Some libraries have more land available to turn into garden beds. The Northern Onondaga Public Library in Cicero, New York, has a half-acre Library Farm. Their patrons can "check out" a garden plot for the season in one half of the area. The other half of the garden is where anyone can work, with no other commitment. The food grown there goes to the food bank and to the garden helpers. A library with even a tiny garden might get the ball rolling to turn other public spaces into gardens. Eventually gardens will be everywhere. There have been communities planned from the beginning with the public areas as food-producing spaces. Can you imagine walking around your neighborhood, picking food wherever you find it? The Village Homes community in Davis, California, would be a model to look to for that.

Not everything to be planted is food. We need all kinds of plants in our ecosystems. My specialty is growing food, but all sorts of seeds

Carnegie Library of Pittsburgh - Main.

Garden behind the sign at the Carnegie Library of Pittsburgh - Main.

should be available in a seed library. In the "Companion Planting" chapter of my book *Grow a Sustainable Diet,* I stress the importance of expanding the diversity of your vegetable garden with flowers and herbs to create a balanced ecosystem. Including fruit is even better. Snacking on strawberries and grapes while you are still in the garden is a terrific experience.

Seed libraries go hand in hand with gardens, but working with the seeds and developing gardens are both huge undertakings and are often developed as separate projects. The seed library at the Mountain View Public Library in California is fortunate there is a garden at the fire station nearby. One of the programs sponsored by the seed library includes a walk to Mountain View Fire Station 1's edible garden. The more that local projects can support each other, the better. Imagine a garden at every fire and police station and every other available space, supported by resources from the library in the form of seeds, educational material, and programs! And then there is the cooking. If food is being produced, it will be consumed, and libraries generally have a good cookbook section. There are just so many things to know about, and one thing leads to the next.

Gathering Together

Libraries often sponsor book clubs, acquiring many copies of the same title so a group can read the book at the same time. Although a book on seed saving might not be something a group would want to read cover to cover, there are many books on related issues (sustainability, biology, etc.) that would interest a book club with plants and seeds as their focus. So often, people want to do things, but they don't know how to go about getting started. Reading about plants and getting to know other people with similar interests is a great first step, especially for those whose garden ambitions are currently landless.

A library could host a seed club, or combine it with a garden club for the express purpose of learning more about and maintaining the seed library. It doesn't even have to be a club — just a monthly meet-up at the library. A local food group could also meet at the library, taking an interest in the seeds being offered, along with the cookbooks. Lining up speakers for presentations can get expensive, but groups don't always need an outside speaker. Using the resources already available at the library, they can learn from each other. When people are passionate about something, they are usually excited to share what they've learned. That said, it is good to have a budget to pay speakers now and again. People who have devoted much time and energy studying and working in an area to gain expertise should be compensated for their time, travel, and knowledge when giving a program, and honorariums are always appreciated.

Declaring Seed Independence in Public

Gandhi said that one way for the Indian people to become free of British rule was to spin their own cotton. Cotton was a major product of India, however the Indians received a low price for it when it was sent to Britain to be spun and woven into cloth. The Indian people had to pay a high price to get it back as cloth. Developing spinning and weaving industries in India would be an important step toward independence. Cloth produced by the people of India is called khadi. Wearing khadi still carries meaning today. Gandhi wasn't just satisfied that the people take control of their lives by spinning cotton; he said it was necessary to spin it *in public.* There was a contest to encourage the design of a new spinning wheel that was small enough to be used anywhere. *Charka* is the name for spinning wheel in India. The design that resulted from the contest is for a charka that can fold up to make a box the size of a book. It is called a *book charka.* Taking control of their lives and spinning their own fiber was such an important part of Indian independence that the Indian flag carries a circle that represents a charka.

Gandhi wasn't the first to promote spinning in public as a means of non-violent demonstration. During the time of the American Revolution, women in the Colonies would gather with their spinning wheels to work with wool and flax to show their support for boycotting British goods. Tea was not the only thing being rejected by the people seeking independence. Wearing homespun was an act of patriotism, just as wearing khadi in India was.

The danger of the takeover of our food and our lives by multinational corporations is on the minds of many, and for good reason. After

Amanda West saw the movie *The Future of Food,* she was angry and dis-illusioned. She wanted to help make the world a better place, but other than demonstrating against Monsanto and expressing her anger, she didn't know any other alternatives — until she learned about seed librar-ies. Seed libraries are a way for us to bypass Monsanto and other such companies. To begin her journey to seed independence, she used seed libraries as her thesis project, resulting in putting seeds in the Pittsburgh libraries. Sharing seeds in public keeps them in the public domain. If you enjoy thinking of it in these terms, just as with those handspinners from years past, it is an act of non-violent protest. I enjoy thinking in terms of how important it is in building a new food system — one where each of us has an essential role. The public libraries are already something we own; we paid for them with our taxes. The reason public libraries exist is to encourage learning, which allows us to take control of our own lives. Public libraries are the ideal institution to have a role in the new community food systems that are being developed everywhere.

CHAPTER 4

Other Entities to Pair With

Despite their name, seed libraries are not always located in libraries. And even when they are physically located in book libraries, seed libraries may be strongly supported by other organizations: Transition Town groups and other community resilience efforts, Permaculture groups, student groups, land trusts, and Extension Master Gardeners. The actual physical location of a seed library might be in a museum or anywhere people gather. A seed library might have its start in one location and move to another after a year or two before it settles in one place. If a seed library is to stay viable, it will evolve with the community according to the help that is available to keep it going. I'll address funding in Chapter 6, but if you are trying to get a seed library started and will be applying for grants or other funding, partnering with a nonprofit organization already in existence is an advantage if you do not have nonprofit status.

Transition Movement

The Transition Movement developed as a result of the work of permaculture educator Rob Hopkins. Permaculture is a design system whereby all the energies within a system are used to maximum efficiency;

the excess from one operation becomes a resource for another. The Transition Movement applies permaculture to the whole community. Communities instituting initiatives following these ideas are often referred to as *Transition Towns*. Hopkins wrote *The Transition Handbook: From Oil Dependency to Local Resilience* (2008) and *The Transition Companion: Making Your Community More Resilient in Uncertain Times* (2011). His most recent work, *The Power of Just Doing Stuff: How Local Action Can Change the World,* was published in 2013. The Transition Movement is about transitioning to a way of life that is less dependent on fossil fuel and more dependent on people helping their communities move forward in these changing times. Two words I keep seeing in my research into Transition activities are *community* and *creativity.* Rather than turn their backs on society as it is now, to build something new, Transition groups work from where we are to make it better.

The work of a Transition group can take many forms. Participants volunteer for committees and projects according to their interests. Beginning with Rob Hopkins' projects, first in Kinsdale, Ireland, in 2005, then Totnes, England, in 2006, community projects have grown to include 1,107 Transition initiatives in 43 countries by 2013. Find out more about the Transition Movement at www.transitionnetwork. org.[1] Information specific to the United States can be found at www. transitionus.org.[2] The website www.transitionnetwork.org/initiatives/ map is a map that shows both official initiatives and mulling groups (they're mulling it over — deciding if they are going to become official or not). You may find a group near you or be inspired to start one.

One well-known seed library that is connected with a Transition Initiative is the Richmond Grows Seed Lending Library in Richmond, California. The Richmond Rivets is the Transition group behind this seed library which is housed in the Richmond Public Library. Rebecca Newburn, a permaculturalist and teacher, is a charter member of the Richmond Rivets and co-founder and coordinator of the seed library. Besides the library, the Richmond Rivets sponsors a weekly crop swap in two locations during the summer months to share food, plants, recipes, and gardening tips, and to build community.

Richmond Grows Seed Lending Library

Richmond Public Library-Main branch
Richmond, California

Began in 2010

Mission: Our Mission is to increase the capacity of our community to feed itself wholesome food by being an accessible and free source of locally adapted plant seeds, supplied and cultivated by and for Richmond area residents. *Richmond Grows* celebrates biodiversity through the time-honored tradition of seed saving, nurtures locally adapted plant varieties, and fosters community resilience, self-reliance, and a culture of sharing. We celebrate our human diversity through outreach and inclusion. *Richmond Grows* strives to fulfill its mission by focusing on two activities:

1. To establish and grow a seed library — a depository of seeds held in trust for the members of that library — available to all Richmond residents;
2. To provide information, instruction and education about sustainable organic gardening.

The Richmond Grows Seed Lending Library is a project of the Richmond Rivets, a Transition Initiative, in collaboration with the Richmond Public Library. It is a fiscally sponsored project of Urban Tilth, an organization dedicated to cultivating agriculture in west Contra Costa County, California. Co-founders are Rebecca Newburn and Catalin Kaser.

Printed materials are in English and Spanish. They welcome volunteer translators, especially for Laotian and Mien.

Website: www.richmondgrowsseeds.org. Includes an orientation video in both English and Spanish.

Rebecca Newburn maintains www.seedlibraries.net, a website where seed libraries can find and share resources. There you will find a list of Sister Seed Libraries and information about starting a seed library. Videos from other seed libraries are posted there, also.

She also moderates www.SeedLibraries.org, a social network launched by Devon Grissim, Elan Goldbart, and Andrew Whitman.

Permaculture Organizations

Permaculture work is where Rob Hopkins began with his Transition Towns. If you are serious about the study of permaculture, you know it includes all aspects of our daily lives, with sustainable food growing being one part. What kinds of seeds are used to grow our food and where they come from are the foundations of a sustainable food supply. The main work of many permaculture organizations is to sponsor the 72-Hour Permaculture Design Certificate Course. Other opportunities, both short (hour-long lectures) and long term (apprenticeships) can be found through Permaculture groups. The Blue Ridge Permaculture Network in central Virginia regularly publishes an online newsletter of upcoming events in the region related to permaculture. Some Permaculture groups, however, are formed to "spread the word" and teach the skills of permaculture without hosting certification classes.

At the Central Rocky Mountain Permaculture Institute (CRMPI) in Colorado, Stephanie Syson was the driving force behind the Basalt Seed Library located in the Basalt Public Library. The Roaring Fork Food Policy Council was a partner in establishing the Basalt Seed Library in 2013. CRMPI offers training in permaculture design. In addition, you will find hands-on demonstration site tours, an on-site edible landscaping nursery, and weekend workshops there. Under Syson's guidance, the seed library has spun off from CRMPI with the town of Basalt stepping in to establish a seed saving garden and a Permaculture Food Forest in a Basalt public park. The seed garden will help supply the seed library with seeds. As seed libraries establish themselves, transitions in sponsors and management are inevitable.

In California, the Chico Seed Lending Library (CSLL) started in 2013 at the Chico Branch of the Butte County Library as a collaborative project between the Butte County Library, Chico Permaculture Guild, GRUB Education Program, and other community partners. Its mission is to preserve genetic diversity and food sovereignty, to collectively develop locally resilient plants and to offer free access to seeds for growing food, flowers, herbs and native plants. CSLL's goals include the expansion of community education in seed saving, gardening, permaculture, and current topics and issues in agriculture. Community-building is also an important aspect of this project. Founded in 2009, the Chico Permaculture Guild meets monthly at the Chico Branch Library with a potluck meal. Members seek to inspire and help individuals, families, and farms achieve abundance and a lifestyle in harmony with natural systems by providing informative meetings and events, including occasional permaculture work parties called Permablitz events. An annual seed swap was part of the Chico Permaculture Guild's activities for many years before the seed library was established. This permaculture group partners with GRUB on many projects, including the Chico Seed Lending Library and Seed Swaps. GRUB (which stands for Growing Resourcefully Uniting Bellies) is a nonprofit group that puts its energy toward educating children about gardening and nutrition, establishing and supporting school and community gardens, offering workshops, composting restaurant waste, and other related endeavors. The Butte County Library provides the space for the seed library as part of its mission to serve the informational, educational, and recreational needs and interests of Butte County residents.

Student Groups, Citizen Groups, and Land Trusts

College students are the driving force behind some seed libraries. In Chapter 3, I told you about how the Pittsburgh Seed and Story Library (now the Carnegie Library of Pittsburgh-Main Seed Library) was started as a graduate student's thesis project. Another project started by one person is the Sylva Sprouts Seed Lending Library in Sylva, North Carolina. Jenny McPherson is the person behind this seed library. Although she was a graduate student in library science at North Carolina Central University at the time, her work with the seed library was independent of her graduate studies. Her position as the manager of the Jackson County Farmers Market gives her an opportunity to have the seeds at the market each week. The seeds are stored in a cabinet that lives at the Jackson County Cooperative Extension office when they are not with her at the farmers market. Funding and encouragement for this seed library were supplied by the North Carolina Cooperative Extension Service, the Appalachian Sustainable Agriculture Project, the Land Trust for the Little Tennessee, and the Jackson County Farmers Market.

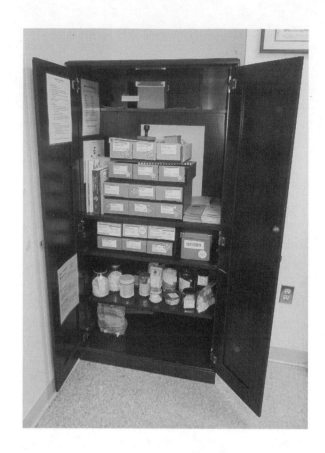

*Sylva Sprouts Seed
Lending Library
cabinet.*

Although it might be necessary to get things started in this way, it is difficult when the seed project depends on one person. The more volunteers, both serious and not-so-serious, that can be recruited to help, the better. In Canada, seed saving became a part of a bigger movement in Toronto. The Toronto Seed Library in Toronto, Ontario, had its beginnings in November 2012 due to the efforts of Occupy Gardens Toronto and students from the University of Toronto and York University. It is a growing cooperative of individuals and organizations, seed savers, gardeners, farmers, educators, librarians, policy makers, and food lovers. The Toronto Seed Library has many branches around the city including locations in two tool libraries (one of which is a Makerspace), a nature center, two universities, a church, a video broadcast business, and Permaculture GTA (Greater Toronto Area) headquarters.[3] In addition, there are traveling branches. This seed library has minimal structure, with each branch fine-tuning the procedures according to their circumstances. In order for as many people to be reached as possible, outreach at events is a part of the Toronto Seed Library. More branches in many different venues, including the public library, are envisioned for the future. It is a labor of love for the volunteers who used their own resources to get it started. More volunteers are welcome for numerous tasks, including translating their materials into other languages. Grimsby Grows Seed Lending Library, located in the Grimsby Public Library, was the first seed library to open in Ontario, serving as inspiration for others in the province to follow.

The Demeter Seed Saving Project at the University of California at Santa Cruz (UCSC) was started in 2011 by student Andrew Whitman to preserve and promote locally adapted organic varieties of heirloom seeds of the Central Coast, and decreasing the community's reliance on seed companies, especially those that were using genetic modification techniques. This project is a student-run, nonprofit organization of local farmers and gardeners working to preserve the genetic heritage of their food. A grant from the Strauss Foundation and support from Measure 43, UCSC's Sustainable Food, Health and Wellness Initiative, helped launch this project. Some seed grow-outs are done at the college farm, and seed swaps are held quarterly (more about seed swaps in Chapter 10). Recipients of the seeds are encouraged to save seed for themselves and to donate back to the library, thereby encouraging greater independence and resilience of both the individual and the organization. This project is currently run by students involved in UCSC's Food Systems Working Group.

Seed libraries may be the result of citizens coming together for a better life in their communities, without a connection to a larger umbrella organization. The goal of the Seed-to-Seed Library located in Fairfield, Connecticut, is for every household to have a garden. This seed library is housed in the Fairfield Woods branch of the Fairfield Public Library and opened in 2011 in partnership with the Fairfield Organic Teaching Farm, which is not actually a farm, but a nonprofit organization led by a group of Fairfield citizens to offer interactive programs to the community. The public library now has a children's garden on the library grounds and a bed in the community garden.

The Lopez Community Land Trust (LCLT),[4] located on Lopez Island, Washington, is a nonprofit organization devoted to building a diverse, sustainable community through affordable housing, sustainable agriculture, and other dynamic rural development programs. A seed library was begun in 2012 and is housed next to the LCLT office. The LCLT Seed Library is committed to increasing the capacity of the local food system by providing island-appropriate open source seeds, while fostering community resilience, self-reliance, and a culture of sharing. It is part of the Sustainable Agriculture and Rural Development (SARD) program of the land trust. Other projects of SARD include an island-grown farmers cooperative, a mobile processing unit to handle cows, pigs, and sheep, and co-development of a school garden and farm program.

Museums and Elsewhere

Keep an open mind when considering where to have a seed library. What you are looking for is a place where people already come, especially people who are interested in gardening. If you want to get seeds back (which may not always be a requirement), you want to have a location where patrons would be coming back anyway. That's what makes public libraries such an ideal place for seed libraries. Tourist destinations may not be the best choices because people who visit there are just passing through. On the other hand, museums that are involved with the surrounding community might be a good choice. One such museum is the Jane Addams Hull-House Museum.

The Hull-House Seed Library[5] opened in 2011 at the Jane Addams Hull-House Museum located on the campus of the University of Illinois at Chicago. This museum is in a settlement house founded in 1889 to provide needed services to poor immigrants. Nutrition and food security were a priority in this outreach to lead to more peaceful communities. Programs promoting nutrition, food security, and peaceful communities should be priorities everywhere in this day and age, and promoting gardening and seed saving are appropriate activities to include. A "Sparks! Ignition" grant from the Institute of Museum and Library Services was instrumental in establishing this seed library. The seed library continues the mission of the founders.

The Jewish Community Center in San Francisco became home to the JCCSF Rooftop Garden Seed Library in 2013. The garden program there seeks to intertwine local produce with the Jewish value of "repairing the world," including caring for the environment. Seeds are an ideal way to preserve your heritage and pass it on to the next generation. Wherever there is a connection, whether it is social, cultural, or ethnic, food generally plays a part. Gardens that provide produce for those special dishes deepen the experience. Making seed saving a part of a garden is a natural progression to bring a project full circle.

Cooperative Extension Master Gardeners

When I was a teenager in the 1960s, the Cooperative Extension Service in our county was part of my life. County Extension offices usually had an agricultural agent, a home economics agent, and a 4-H agent. I was active in 4-H. The role of the Cooperative Extension Service in each state was to help farmers and homemakers be more productive. As society changed, so did the Extension Service. Home Economics gave way to Family and Consumer Sciences. The agricultural aspect of Extension transitioned to include information about home horticulture. The Cooperative Extension Service was the place to go for information on any of these topics, backed by the land grant university in each state. To better serve the people in each community in the area of horticulture, states began to implement Master Gardener programs; the first one was in Washington state, in 1972. Master Gardeners receive horticulture training from Extension agents and industry experts. They are trained to be volunteers who disseminate information from the land grant colleges to the public.

Volunteerism is key to being a Master Gardener. In Virginia, a prospective Master Gardener has to attend 50 hours of classroom instruction and then log 50 volunteer hours within a year. After that, to retain their status as a Virginia Master Gardener, they must put in 20 volunteer hours plus 8 hours of recertification training each year. Seed libraries fall within the category of projects that Master Gardeners can volunteer for. Keep in mind, the classroom instruction a person receives to become a Master Gardener covers a broad spectrum of horticulture topics. Unless a Master Gardener has taken the initiative to study a certain topic individually, he or she will only get an overview of the various subjects covered in training, if that. Master Gardeners are trained to find the answers to your questions from all the information available from the Extension Service, not know the answer off the top of their heads. Too often, people expect that if a Master Gardener is working on a project they know everything about it. I have known Master Gardeners who have volunteered with school gardens with no vegetable gardening experience at all, but have a willingness to help. However, they might be the local expert on bulbs, roses, or ornamental shrubs.

In the mid-1990s I was a parent volunteer for the garden program at my children's elementary school. Each year I recruited Master Gardeners for whatever projects we needed help with, knowing they were anxious to get their volunteer hours in. Because I was well known to the people at the Extension office, my phone number seemed to be given out freely. I would receive phone calls on the topics of school gardens, organic gardening, managing worm bins, and hatching eggs. These were all topics not covered in the Master Gardener manual. Finally, one year I gave a workshop at my home on making a worm bin and encouraged Extension to send a Master Gardener to attend so that someone else would gain expertise. A Master Gardener did come. She took home her worm bin and began giving programs on worms, even decorating her bin to look like a circus wagon. Eventually, more school gardens were established and more Master Gardeners became proficient in the ins-and-outs of school gardens.

If you find a Master Gardener who wants to start a seed library, it will be one of those pioneering souls who have educated themselves in the process. Educating others is a major part of what Master Gardeners do. Some enjoy taking the part of worker bees, helping out wherever help is needed, but not taking the lead. Others are born leaders. They are comfortable getting up in front of groups and speaking in public. Seed libraries are the ideal forum for both kinds of volunteers. Besides getting a seed library up and running, there will be a need for ongoing presentations and classes to help the seed savers. Master Gardeners who have had experience with seed saving themselves could handle that job. Ideally, if a Master Gardener takes a seed library on as a project, it will become a project within the county Master Gardener program and involve more than one dedicated individual. I doubt that the particular Master Gardener who carried her worm bin to the schools is still doing that after all these years, but worms became a part of the county program. So, if the Master Gardeners in your county don't have seed libraries or seed saving on their agenda, that doesn't mean it has to always be that way.

Master Gardeners in Perry County, Illinois, worked with the Du Quoin Public Library to start a seed library there in 2011 after one of the Master Gardeners and the librarian both read the same article in *Organic Gardening* magazine about seed libraries. It only takes an idea and a willingness to do something different to get things started. In 2013, the Perry County Master Gardeners received a Teamwork Award from the Illinois Master Gardeners, recognizing their work.

In Canada, there are Master Gardener organizations in Alberta, Manitoba, British Columbia, Ontario, Saskatchewan, and Atlantic Canada.[6] In the UK, Master Gardeners are trained and supported by Grow Organic, UK's leading organic growing charity. These UK Master Gardeners promote local food growing and sharing across the UK.[7] Organization and training of the Master Gardeners in Canada and the UK may differ from that in the United States.

CHAPTER 5

Seeds

S EEDS ARE BASIC TO LIFE. They have the potential to not only grow into food, flowers, bushes, and trees, but to reproduce themselves abundantly. Some cultures hold them sacred, as all cultures should. If we don't value our seeds, we don't value all of life surrounding us. Seeds connect us with our past and with our future. City dwellers may find it hard to imagine just how important seeds are if they never see the food they eat or the flowers they bring home actually growing in the ground and producing seeds. The closest many people come to confronting seeds is when they clean out a pumpkin for Halloween. If that is your experience, take a moment to hold those seeds in your hands and picture the people who once depended on winter squash as a major food source. Think about how they would clean the seeds and store them until the next growing season. Notice the difference in the seeds themselves. You could sort the seeds according to their size and plumpness, then do a germination test on each group of seeds. The results of those tests would teach you what to look for when saving seeds of not only squash, but other crops as well. (More about germination tests later in this chapter.) The book *Buffalo Bird Woman's Garden*[1] is good to read if you are pondering pumpkin seeds. It relates the agriculture of the Hidatsa Indians, featuring sunflowers, corn, beans, and squash; the story is told by an elderly Hidatsa woman who kept to as many of the old ways as she could later in life.

When native lands were taken over by newcomers, one of the ways of controlling the natives was to control their food. This is still going on today, but now it is the corporations who are controlling the food — YOUR FOOD — but only if you let them. We are all eaters. Even if you never grow anything, what you choose to eat determines whether you continue on the corporate-controlled route, or you get back to your roots. If you have always lived with a food supply brought to you by the governing corporations and don't recognize any roots or heritage food, now is the time to change that. When you plant seeds, particularly when they are given to you by someone else, you are putting down both kinds of roots. Even if you move, you can take them with you — both the seeds and the memories. If you move to a different climate, the old seeds may or may not do well. Look at it as an opportunity to find out what grows well in the new place and to make new friends to share seeds with.

In our melting pot of a nation, even before the boats first floated to the shores, people have relied on the seeds that they grew up with. Unfortunately, when the newcomers came to America, they didn't recognize the importance of the indigenous crops that the natives were growing. They didn't recognize the importance of the natives, either, and pushed them off their ancestral lands. If you are not a grower but are searching for a root heritage, buy your food from growers who understand this and have established their own roots. They will bring you into the fold.

Where to Find Seeds

When you start a seed library, you will need a quantity of seeds to get the ball rolling. If you are patient and plan far enough ahead when setting up your project, you can contact gardeners and farmers at the beginning of a growing season and ask them to save seeds for you. That way, you will be starting with varieties that grow well in your region. I wrote about the importance of saving seeds that do well in your particular area in Chapter 2. I will not be telling you how to save seeds, since there are already many books and web resources available that do that. You will find some of them listed in the "Resources" chapter. Have as many of these resources available to your potential seed growers as possible. Make sure the donated seed is not designated as PVP or protected by a utility patent (I explained PVP and utility patents in Chapter 1). Seed catalogs usually have that information in their variety descriptions.

If your project has been funded, you could buy seeds, especially if there are specific varieties you want to acquire. However, at the end of each year, seed companies have seeds left over that they cannot sell the next year. They are still good, at least for a while, but being another year older, they may not meet the standards for sale by the seed company, and the company might be willing to give them to your seed project for free. Some companies ask you to pay the shipping and some don't. With the recent surge in seed libraries, community gardens, and school gardens, seed companies can't always keep up with the requests for free seeds. Keep in mind that, as much as they would like to be charitable, they are businesses that need to make a profit to keep going. Be respectful of that and don't depend on their handouts to keep your seed library going after the first year.

You should acquire seeds from companies that have signed the Safe Seed Pledge. The Council for Responsible Genetics maintains a list of such companies on its website.[2] In order for people to save seeds to bring back, you need to avoid hybrid seeds. In Chapter 1, I explained the difference between hybrid and open pollinated seeds. The seeds saved from a hybrid plant won't necessarily grow out to be the same as the parent. You can get some interesting things from hybrids, but you need something you can depend on for a seed sharing project. You need to have seeds that are open pollinated and not genetically modified.

Care of Seeds

When acquiring seeds, pay close attention to the source. Just like with people, the health of the parent plants affects their offspring. The best seeds need to be saved from healthy, robust plants. When the plants are forming seeds, the weather, the amount of nutrients available to the plants, and competition from weeds all affect the health of the seeds.

It is extremely important to remember that seeds are *alive.* Even if they appear dormant, they are still respiring, giving off oxygen and taking in carbon dioxide. The conditions you keep them in will determine how long they stay viable (able to grow when the conditions are right). Seeds need to be kept cool and dry. Heat is usually detrimental to seeds. Ideally, seeds should be stored at 32–41°F (0–5°C). If you are storing seeds at room temperature, consider the guideline that for every 10°F (5.6°C) the storage temperature is lowered, the length of time the seeds will stay viable is doubled.[3] You could store seeds in a freezer, but you need to make sure they are very dry. Too much moisture in the seeds could rupture their cells when they freeze. Very dry seeds can withstand extremes of heat and cold better than seeds with more moisture. For every one percent increase in seed moisture, longevity decreases by half.[4] A guideline to use is to add the relative humidity of the storage area and the temperature (°F). Ideally, those two numbers should

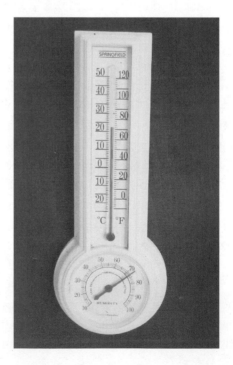

Inexpensive
thermometer/hygrometer.

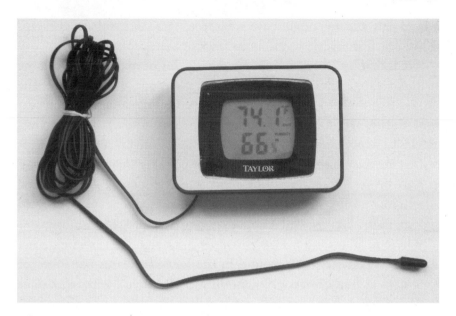

Digital thermometer/hygrometer with remote sensor.

add up to less than 100. In reality, I don't believe my seed storage areas have ever met those criteria. You can monitor your seed storage areas with a combination thermometer/hygrometer.

Hot humid climates are hard on seeds. Nevertheless, even though I live in Virginia which is hot and humid, I don't take extreme measures to save seeds. I dry them at room temperature, leaving them set out on plates or in bowls in the house, stirring occasionally. We do not have air conditioning. After a few weeks, or when I think they are sufficiently dry (timing depends on the seeds and the atmosphere), I store them in airtight containers, usually glass canning jars. Glass jars with lids with rubber seals, such as canning jars, are the best. You can store seeds in paper envelopes and put many envelopes in the same jar. Always clearly label everything, keeping different lots of seeds separate if you think it will be necessary to know the difference in lots later. Seeds that came from different lots of parent seeds, or were grown, harvested, or stored differently might behave slightly different when grown out, or the germination rate might differ. Particularly if you have some seeds that you suspect might not be as viable as others, keep them separate and label accordingly. Or, you might have seeds from some plants that did exceptionally well and you want to keep track of their progeny.

Seeds that have been stored in a refrigerator or freezer need to come to room temperature before the packages are opened; otherwise, the moisture in the air will quickly be absorbed by the seeds. For this reason, seeds stored in bulk in a library for patrons to package themselves should be kept at room temperature. The bulk containers could be kept out during the months they are needed for planting and stored in a cooler place the rest of the year. If the desire is to keep a seed cabinet stocked all year for patrons, in the off-months a limited number of packets could be kept in the cabinet with a note that more is available in storage.

I have to admit, since I *succession plant* through the gardening year, I tend to let my seeds sit around in the house in places not ideal for long-term viability, but convenient for me. However, I know that since I save many varieties each year, my supplies are replenished often, so it is not a problem. A seed library, though, should hold itself to the highest standards it can (although they may not be the highest standards possible), since the patrons are depending on the library for good seeds. If your seed project is in a public building, the relatively cool, dry conditions that generally exist will be fine for most seeds for a couple of years. You do what you can with what you have. From the beginning of your project, it is good to label the seeds with the year they were grown by the donor or packaged for sale by the seed company. A seed library should strive to have a regular turnover of seeds, which may take a few years to establish. A seed bank, on the other hand, stores seeds for the long term and not everything is grown out every year or two. A seed bank would need more strenuous regulation of their storage facilities than a seed library. Books on seed saving and Internet resources have charts[5] available that will give you an idea as to the longevity of the seeds of each crop, but the best way to tell if they are still viable is to do a germination test.

Germination Test

Doing a germination test is a way to predict how well your seeds will sprout in the garden. If you have ever sprouted seeds to eat, the concept

Mississippi Silver cowpeas with 80 percent germination. Two days later, they achieved 100 percent germination.

is the same. Keep the seeds moist and they will grow. With a test, you need to be able to count the number of sprouted seeds to determine the percent of germination. Most directions suggest using paper towels to do a test, but I prefer using a coffee filter, so that's what I'll refer to. With a pen (no water-soluble ink), write the name of the variety and the date on the coffee filter, along with any other information that you want to remember about the seeds. Thoroughly wet it, then press out the excess water. Place a minimum of ten seeds on the filter. The more you put there, the more accurate your test is, but make sure to record the number. Seed companies use 100 seeds at a time in their tests. Fold up the damp filter with the seeds inside and put it in a container with a lid to keep it moist. Numerous tests can be put in the container at the same time. After a few days, take a look at what is happening. If 8 of the 10 seeds germinated, you have an 80 percent germination rate. If you started 20 seeds, 80 percent germination would have resulted in 16 seeds germinating.

Some seeds take longer to germinate than others, and you will have to put the folded filters back in the container and check again later — adding some water if necessary. Be patient. Pepper seeds, for example, can take as long as three weeks to germinate. When you are sure all that is going to germinate has done so, record the count. Label your supply of those seeds with the date and germination rate. When seeds are donated from a seed company, they may have the date and germination rate from the last test on the packaging. Sometimes they are labeled as having poor germination. When you distribute those seeds, make sure your patrons know that. I once acquired some cabbage seeds that were rated as poor germination in a seed swap, and, in fact, nothing grew. It was good to know from the beginning that the problem was with the seed and not something else. Poor germination may be due to the age

Seed packet showing 98% germination from a test done in March 2013.

of the seeds; however, it can also be due to other factors. Some seeds require *scarification* to help crack the seed coat; other seeds need light to germinate. Research the requirements of the specific crops you are working with.

There is a minimum legal germination rate that seeds have to meet before they can be sold. The seed projects involved with this book don't involve sales of seeds, but it is good to know what those rates are — and how different the rates can be for some crops. For example, it might be interesting to know that the minimum legal germination rate of cucumbers and lettuce is 80 percent, tomatoes 75 percent, and carrots 55 percent. You can find the minimum legal germination rates for seeds in the Master Charts in *How to Grow More Vegetables*[6] by John Jeavons. The germination rate needs to be taken into account when distributing seeds to your patrons, whether they are distributed pre-packaged or the patrons help themselves. Not every seed can be counted on to grow. However, with good, fresh homegrown seed, you just might have some with 100 percent germination.

Plant Population Numbers

For the most genetic diversity, you would want to save seeds from as many plants as possible, not only the best ones. Plants with genes that allow them to do well with less water will do the best in the dry years. Conversely, plants with genes that allow them to thrive with more water will do well in wet years. And so it goes for other characteristics. You want to preserve as diverse a variety of genes as possible because each year is different. Save from the best plants, but also save from a cross section of plants to preserve the genes that express themselves the best under conditions that are not present that particular year. There will likely be some *rouging* (taking out anything that is not true-to-type), so that needs to be allowed for when deciding how many seeds to plant. Familiarize yourself with the common characteristics of each variety you are working with so you know what true-to-type is. The descriptions and photos in the seed catalogs will guide you.

How each crop produces seeds affects the minimum number of plants necessary for good diversity. With self-pollinating plants (also known as *selfers*), such as peas, beans, and tomatoes, 20 plants are enough to save seeds from if you are trying to preserve the whole variety. If you save from fewer or only one plant, you will be saving a sub-line of that variety, otherwise known as a *strain*. For plants that are pollinated with the help of the wind or insects, such as corn, kale, and sunflowers, 40 to 200 plants are recommended to maintain the variety. Most home gardeners save from far fewer plants than that each year. But if your seed library has more than one gardener bringing back seeds of the same variety, those seeds combined could make the necessary number of plants saved from. I knew a gardener once who saved his own squash seeds, but in some years, he would buy seeds to grow along with his saved seeds. He said it added diversity to his squash. On the other hand, if you are developing a local strain with certain characteristics unique to your area, this seed-breeding effort might begin with only a few special plants.[7] You can learn more about that from *Breed Your Own Vegetable Varieties* by Carol Deppe. The danger of cross pollination and isolation distances — how far apart to plant to prevent cross pollination — is found in most seed saving resources, and this is information you want to make sure your seed savers have.

Seed Saver Organizations

Fortunately, some forward-thinking folks have recognized the necessity of saving seeds and have established organizations to do so. Sometimes, rather than having a plan, they were simply following their hearts. In Chapter 1, I briefly introduced you to Seed Savers Exchange. The gift of family seed and the knowledge that they were the only ones concerned with preserving it sparked an interest in Diane and Kent Whealy to connect with other seed savers. In an effort to trade seed with like-minded folks, they began the True Seed Exchange in 1975 with 29 members and a six-page listing of seeds and the gardeners willing to share them.

The name was changed to Seed Savers Exchange (SSE) in 1979, and the listing of seeds available from seed savers became the *Seed Savers Exchange Yearbook,* available to all members. Through the annual Yearbook, gardeners and farmers with no family member interested in taking over their seed collections finally had a place to share their seeds with people who would grow them. Gardeners now had access to heirloom seeds that had a connection to a person and place. In those days before email, True Seed Exchange/Seed Savers Exchange received many hand-written letters from gardeners telling their stories, with or without the seeds themselves. Although the Whealys' intent was to facilitate exchange among growers, people began sending them seeds. In at least one instance, they were able to unite a woman with beans descended from the ones her grandmother had contributed 15 years earlier.

Well aware of the loss of varieties and of seed companies, in 1984 SSE began acquiring endangered varieties to maintain in their collection, eventually establishing a practice of seeking varieties that were only offered by one company. That project was assisted by Kent Whealy's work compiling a list of all mail-order seed companies in the US and Canada and the non-hybrid varieties they carried. This became the *Garden Seed Inventory* and is a resource for finding varieties offered by only a few sources. The sixth edition of the *Garden Seed Inventory* was published in 2004. When the final source stops carrying a variety, for whatever reason, it is just *gone* unless someone, such as a private seed saver, has been working with it. If you are getting into seed saving and want to make a difference, you might follow the Whealys' example. You could choose a variety offered by only one source, grow it out to seed, and share it with others through your seed library.

In 1992, Seed Savers Exchange began offering seeds for sale. Anyone could purchase them, even non-members. Today Seed Savers Exchange is run by a board of directors and there are demonstration gardens, a visitors center, heritage breeds, and an annual gathering at Heritage Farm, its headquarters in Decorah, Iowa. You can follow their work through their catalog and website.[8] You will find seed saving tips in the catalog. Become a member and you will receive the Yearbook. Anyone can view the Yearbook online to see what is available, but you have to be a member for total access. Through their Herman's Garden Seed Donation Program, Seed Savers Exchange offers seeds to established community and educational groups who promise to freely share the harvest and save seed for others in need. The donated seeds are those returned from retail outlets or are overstock seed packets from the previous year.

Seeds of Diversity Canada, also briefly introduced in Chapter 1, had quite different beginnings. In 1984, the Canadian Organic Growers held a conference with Kent Whealy as a keynote speaker. The Heritage Seed Program developed from that, under the umbrella of the Canadian Organic Growers. The Heritage Seed Program eventually changed its name to Seeds of Diversity Canada, becoming an independent charitable corporation operated by a volunteer board of directors. It is bilingual, with Semences du Patrimoine as its French name. A list of seed swaps in Canada is maintained on its website. An annual *Member Seed Directory* is published following the same format used by Seed Savers Exchange with their Yearbook. When accessing seeds from either the Canadian or Seed Savers Exchange directories, growers are encouraged to save seeds to exchange with others after growing out the ones they receive.

One of the projects currently underway by Seeds of Diversity Canada is the Canadian Seed Library, which is a collection of seeds that backs up the work of member seed savers and Canadian heritage seed companies. Samples of Canada's rarest seeds (and some not so rare) are stored in low-humidity freezers to keep them viable and available for future gardeners and farmers. Another Seeds of Diversity project is Pollination Canada, a project that encourages conservation of native pollinators. In 2008, Seeds of Diversity held a meeting with Tom Stearns of High Mowing Seeds (Vermont) as speaker. That was the catalyst to form the Eastern Canadian Organic Seed Growers Network, an independent association geared toward helping growers to produce top-quality certified organic seeds in Canada. Seed saving organizations have been formed around the world, including the Heritage Seed Library in the UK,[9] the Irish Seed Savers Association in Ireland[10] and the Seed Savers Network in Australia.[11]

One organization that is truly regional is Native Seeds/SEARCH (NS/S), a nonprofit seed conservation organization based in Tucson, Arizona. It was founded in 1983 by Gary Nabhan, Mahina Drees, Barney Burns, and Karen Reichardt. The focus of NS/S is to preserve genetic diversity in crops from the southwestern US and northwestern Mexico regions by establishing a seed bank; over half of the accessions are corn, beans, and squash. This is where freezing seed for longevity is important. Grow-outs to regenerate the seed for the seed bank are done at the NS/S Conservation Farm in Patagonia, Arizona. More about seed banks in Chapter 10.

Native Seeds/SEARCH has a seed grant program for organizations working on educational, food security, or community development projects in the regions where their seeds are best adapted. NS/S offers free membership and limited quantities of free seeds to Native peoples living in the Greater Southwest region. NS/S accepts memberships to support their work, has a seed catalog, and operates a retail store in Tucson, Arizona. In early 2012, a seed library was added to the retail store.

Seed School

Several of the seed libraries I researched mentioned their seed librarian had been to Seed School. Seed School began in 2010 under the direction of Bill McDorman and Belle Starr in conjunction with their seed business, Seeds Trust, in Cornville, Arizona. Their ultimate goal was to build a broad network of regional seed systems. In 2011, they joined Native Seeds/SEARCH as executive directors and brought Seed School with them. Seed School continues with McDorman and Starr through the Rocky Mountain Seed Alliance. To quote from the Rocky Mountain Seed Alliance website: "Students walk away from this innovative learning experience with the knowledge and inspiration to start their own independent seed initiatives, such as community seed libraries and exchanges, seed growers cooperatives, heirloom seed businesses, and participatory plant breeding projects".[12]

The original Seed School was a six-day affair. McDorman and Starr now have a one-day Seed School that they coordinate throughout the Rocky Mountain West; they also accept invitations to conduct the one-day or six-day versions elsewhere. McDorman, Starr, and John Caccia (a graduate of their 2010 Seed School) formed the Rocky Mountain Seed Alliance[13] in 2014. This nonprofit organization is dedicated to

"In 2013 I had the pleasure of meeting Bill McDorman and Belle Starr while visiting Seed School to shoot my documentary *Open Sesame — The Story of Seed.* I witnessed firsthand the incredible passion, in-depth knowledge and enthusiasm for all things seed related that they shared with their students. I watched participants discover not only the world of seeds, but also a new power within themselves. I can attest to that myself having come to shoot a movie and having left with the never-before-considered goal of starting my own seed company."

Sean Kaminsky, Director: *Open Sesame — The Story of Seeds*
www.opensesamemovie.com; Founder: Free Wild Seeds, www.freewildseed.com

strengthening seed and food security in the region. It will help train and support a regional network of community-based seed stewards to grow, store, and distribute seeds for a wide variety of edible vegetables, grains, herbs, native wildflowers, and grasses. Every region should have such an organization. At www.rockymountainseeds.org you will find information about the Seed School and a list of small, bio-regional seed companies started by, or purchased and run by, the students and teachers of Seed School. The Seed Library of Los Angeles (SLOLA) hosted a Seed School in February 2014. David King, founding chair of SLOLA, attended Seed School in 2011.[14]

In Hawai'i, The Kohala Center recognizes that the state imports nearly 90 percent of its food and 99 percent of its seed, and it wants to change that. This independent, not-for-profit, community-based center for research, conservation, and education launched the Hawai'i Public Seed Initiative to help farmers and gardeners select, grow, harvest, sort, and improve seed varieties that thrive in Hawai'i's unique conditions. The first Seed Production Basics for Farmers and Gardeners workshops were held in 2011. The Hawai'i Public Seed Initiative has broadened awareness among growers, farmers, and consumers statewide of the need to save and share the most viable and vulnerable varieties of seeds best suited for Hawai'i's climate and soils. By empowering networks of committed individuals on Hawai'i's five major islands, the Initiative supports Hawai'i's aspiration for greater food production and security.

Finding seeds that are unique to your area and culture will be one of the adventures you will take on when establishing your seed library. Businesses that sell food and other products of seeds need to be concerned with money. Production is top priority for them. Seeds have much more to offer besides monetary profit. The health we come to know from working with seeds and plants and from eating nutritious food grown from them should take top priority for us. Growing plants can trigger all of our senses and bring us much pleasure.

CHAPTER 6

Getting Started

THE FIRST THING I SUGGEST doing to get your seed library going is to involve others. Talk to other people and give them information to read or sources to look into. (Of course, have your potential partners read this book!) Often all it takes is a magazine article or a news broadcast to spark somebody's interest. If your endeavor is a project of a Transition or Permaculture group, you may already have people prepared to move forward with you. If you are a librarian in a public library, help from beyond the library would benefit you greatly. If you are not a librarian and intend for your seed library to be located in a public library, now is the time to bring one on board.

You can get some ideas about who to attract to your project in Chapter 4. Seed libraries could have a committee to support the project. Other names for this group might be team or advisory board — whichever suits you the best. Not all committee members will be the ones to source seed or physically do anything, but they will be the ones who will get the word out to others who need to know. Also, they may be the ones tied to funding. You might find people for your committee already involved in food and nutrition endeavors or community gardens. It is good to have people in the mainstream as well as creative thinkers on your committees; they may approach projects with different views that can be helpful. Someone with an art talent who is willing to share their skills is a plus to have in your group, as well as someone with computer and social media skills.

A representative from your county Master Gardener program can be your link to acquiring volunteers when you need seeds sorted, tested, or packed. The same goes for having a representative from Scouts, 4-H, and any other local organization with a group of potential volunteers. A representative from your local farmers market could be your link to people who are already saving seeds. Religious groups with a penchant for service may like to be involved. Even if you don't establish a formal committee, a gathering of people such as this in your community early on, to let them know of your intentions, would be a step in the right direction.

Mission Statement and Name

You will need a mission statement. This will be the guide for all the actions that follow and serve to let others outside your circle know what you are about. From the beginning, decide if your main goal is to distribute seeds only or to distribute seeds and have the recipients save what they've grown and donate them back. If the latter is the case (which it usually is), you will need to have an education component to your mission. Although many people are gardeners, few may have saved seed before in a way that would be beneficial for your library. If you are only concerned with distributing seeds, you will need to plan for a continual source of seeds. Seed companies have programs to help seed libraries get started, but I can't imagine they would be willing to send you seeds every year.

Maybe your mission is to develop a local resource for seeds specific to your area. There may already be seed savers in your area who would love to donate seeds to your library and possibly come to a program to talk about it. The stories connected to the seeds are an important part of saving our cultural heritage. Some of the local seed savers most likely are getting on up in years, so the sooner you find them, the better chance you have of collecting seed from them to save for your community.

Here is a list of some of the things that seed libraries have included in their mission statements:

- Increase library usage and community involvement (public library)
- Develop a network of skilled seed stewards
- Educate members in ways to save seed
- Reclaim seed as a public resource
- Have safe alternatives to GMOs
- Develop a source of open pollinated seeds that are specific to the locality
- Contribute to and support community gardeners
- Conserve endangered varieties of seed
- Foster a community of resilience and self-reliance
- Support genetic diversity and community sovereignty
- Transmission of knowledge from one generation to another through stories
- Empower members through a deeper connection with nature
- Preserve seed as a sacred trust
- Reflect all cultural diversity of the city
- Serve underserved populations
- Collective action to build a sustainable community food system
- Prevent hunger
- Promote a healthy diet
- Help low income households afford nutritious food
- Restore indigenous varieties of seed

Once you have your mission defined, choose a name. If your project is part of a larger organization, such as a public library, museum, or community group, the name you choose could include the name of that organization to clarify the connection. The name would also include your activity with seeds. Some choices are *seed library, seed lending library, living seed library, seed share,* and *seed exchange.* Knowing what your mission is will help you choose a name.

The Goochland Community Seed Lending Library began as the J. Sargeant Reynolds Community Seed Lending Library at J. Sargeant Reynolds Community College in Goochland, Virginia. A year after the

Goochland Community Seed Lending Library

Reynolds Community College
(formerly J. Sargeant Reynolds Community College)
Goochland, Virginia

Began in 2013

Mission: Goochland Seed Lending Library is a free community seed bank started to establish a local seed supply *for* the community to be maintained *by* the community to encourage the growing of food and the valuable skill of seed saving. It builds community resilience, self-reliance, and a culture of sharing.

This seed library was started by sustainable agriculture instructor Betsy Trice in cooperation with the college library where the seed library is located. The original name was the J. Sargeant Reynolds Community College Seed Lending Library, before the college shortened its name to Reynolds Community College. Trice changed the name of the seed library to Goochland Community Seed Lending Library to reflect the fact that it is open to all members of the community, not just the students and staff at the college.

Although established independently of official classes, Trice's students have the opportunity to participate in germination tests and other duties of managing a seed library. Seed was donated from seed companies to start. Donations now come from patrons and the college gardens.

This seed library is self-serve, with the seed stored in bulk in envelopes and jars. A cabinet designed to store CDs is used for the seed cabinet, with more seed jars and supplies stored on a table where seed catalogs and other print resources are kept. Seed library patrons have access to a wealth of books, DVDs, and Internet resources in the college library. The seed library is accessible whenever the college library is open.

A Facebook page is maintained for the seed library.

seed library was established, the college began the transition to shorten its name to Reynolds Community College. In order to better reflect the fact that the seed library was open to all community members, not just those associated with the college, the name of the seed library was changed to the Goochland Community Seed Lending Library. Don't hesitate to make changes when it is warranted.

Budget and Funding

The amount of money you need to begin your project will depend on how you go about it. If your seeds are donated to begin with, and you quickly develop a group of seed stewards to keep the seeds coming in, seeds are not a cost. On the other hand, you might want to begin your project with certain varieties that you have bought fresh — packed for the current year — to ensure your success. That will be a cost that you can calculate from reading seed catalogs. If you are located in a public library, the supplies for setting up the seed sharing, such as envelopes, notebook, labels, and such (more about that in Chapter 7) could come out of the library's general operating fund. If you are not part of a public library or a similar institution, that cost might be coming from someone's pocket.

How the seeds are stored and distributed will determine how much money is needed to start. Look around for any available cabinets and storage containers; otherwise, you will need to acquire some. A public library may already have books and videos on seed saving, or they may be willing to obtain them through their normal acquisitions. Even if your seed project is not located in a public library, the resources there will still be a valuable asset to educating your members. Acquiring speakers, which is always a good thing, may involve honorariums (unless you know qualified individuals who will speak for free or who need to accrue volunteer hours to fulfill a requirement). Be respectful of someone's hard-earned expertise and plan on honorariums where warranted. Having refreshments at gatherings should be considered in your budget. Whether it is for a launch party to begin your seed library or a lunch party for seed volunteers throughout the year, food is always welcome. Of course, food could be provided by potluck donations brought by the participants for these events.

To provide needed funds, you could solicit donations. It may be easier to obtain monetary donations if the gifts are tax deductible. For that, you would need to be registered as a nonprofit organization or be under the umbrella of one. Personally, I always hate to be involved in activities that ask people for money. Bake sales are good ways to earn some up-front cash. Granted, you are asking people to not only bake something, but also purchase the ingredients to do so, but somehow it translates differently to me than just passing the hat. Maybe if you suggest a bake sale, the non-bakers in the group will say to just pass the hat. People who would not notice you otherwise, will stop and buy something at a bake sale. Bake sales can be combined with yard sales. Yard sales are great because they promote reusing things. Supporters can clean their houses and garages and donate unneeded items to be sold for the cause. If you have a yard sale, also have a plan for what you are going to do with the leftovers. Large thrift stores may send a truck to collect what is left at the end of the day if you schedule that ahead of time. A plant sale to benefit a seed library is also a possibility. Gardeners will both donate and buy plants; plus, you'll be bringing the seed library to their attention. Any of these activities could be combined with a public library's regularly scheduled book sale.

An art and/or a craft show could generate needed funds either by artists and crafters paying for a booth or donating items to be sold for the benefit of the seed library. A silent auction at a potluck event, concert, or whatever other event you think up, could bring in some money. In a silent auction, donated items can be set out with suggested minimum bids on a paper for each entry. People have an opportunity to write down their bids up to the cut-off time. Any activity that involves people coming together is a newsworthy event and good publicity for your seed library.

I'm most familiar with the activities I've just mentioned and less familiar with acquiring funding from grants. If it is grant money you are after, having someone on your committee who has experience working with grants is a definite bonus. Some of the seed libraries that I've researched had grant money to get their project going. One of these grant sources is the US Institute of Museum and Library Services, but there are many others. Begin to notice any projects that list their grant sources, and make a note of them. There are many community groups that would donate to a seed library if its mission fell within the parameters of the mission of the group. You might keep that in mind when crafting your mission statement. Organizations with a gardening or conservation background are likely sources, but with the current concern for health, particularly anything that promotes combating obesity and diabetes, the field expands. Gardening certainly promotes health in the form of exercise, being outside, and eating more vegetables.

A seed library can be tacked onto another project, which is what happened at the Pittsburgh Seed and Story Library. A grant was awarded for a gardening program at the Carnegie Library of Pittsburgh at the same time the seed library was being established.

I have known people who feel they can't start a project without a grant, and through my sustainable agriculture activities I have attended many programs funded by grants. Some great things were being done, but only for as long as the initiative was funded. Grants do tend to bring publicity with them, which is a plus for a project. My reservation about depending on this support is that you can't be sure how long you will have it. Sure, you have the initial award, and if you structure your project with the understanding that you won't need more money to continue, you will be fine. When the economy falters, which it is prone to do, grant money is one of the first things that disappears. If the grant pays an intern to establish the seed library, what are the plans for after the intern leaves?

Years ago, I volunteered on the garden committee at my children's elementary school. The teacher who initiated the project was awarded a $1,000 grant at the beginning. It was spent on tools, educational materials, refreshments when the teachers met for garden meetings, and whatever else was necessary to get the program started. The garden committee had a plant sale every year when the school had its "Spring Fling" and earned enough money to cover ongoing expenses for the garden after that. Most of the plants for the sale were donated by the committee members (myself and a few teachers) and some parents. Many were perennials that we needed to divide anyway. When I left after four years, the garden committee had more than the original grant amount at their disposal, with the plant sale being their only fundraiser.

Define the Space

Space needs to be available for the seeds and for recordkeeping. The seeds may already be in packets for patrons to take. This involves more work to prepare ahead, putting the allotted seeds in each envelope and labeling them. On the other hand, if the seeds are going to be available in bulk, with the patrons helping themselves (hopefully, according to the suggested guidelines that you have developed!), a space for containers, envelopes, labels, and pens needs to be available. Patrons usually record in a notebook what they have taken. Some libraries have a barcode on each seed packet; patrons "check out" their seeds at the main desk, just like books are checked out.

One of the fun things happening at seed libraries is the recycling of now-antiquated card catalogs; many libraries still have them in storage, and they are perfect for holding packets of seed. The Washington County Seed Savers Library in Abingdon, Virginia, received a card catalog from a retired librarian who had it in her basement. It was painted by a local artist. You will find a photo of that cabinet in the color section. Lacking an old card catalog, storage cabinets for CDs work well to hold the seed packets. That is what is used at the Goochland Community Seed Lending Library in Goochland, Virginia. A shelf or two to hold jars of seeds and boxes of seed packets would suffice, also. If the patron needs to fill his or her own seed envelopes and to record in a notebook, a small table and chair are necessary. Books on gardening and seed saving should be nearby. Resource material in the form of brochures, pamphlets, and booklets could be stored in binders or a filing cabinet. A small display of these resources may point the way to the whole collection.

Locate the seeds where it is convenient for both staff and patrons. As with the books, patrons will be browsing the offerings. A location near the door gives visibility to the seed program, but good signage will give it visibility in other locations, as well. The staff should be able to monitor it easily to make sure children are not playing with the

Goochland Community Seed Lending Library.

seeds. Although most seed patrons will abide by the guidelines that
have been established, I have heard stories of the occasional patron who
helped themselves to more than their share of seeds. All library staff
should be trained to keep an eye out to prevent that from happening.
If the seeds are in a small cabinet, it may be located on top of a book-
case already in the library. All in all, the seeds and everything that goes
with them should fit in the space of a computer or study carrel — or
two. Envelopes, labels, and pens can be stored in a drawer of the seed
cabinet.

Website and Social Media

I can't stress enough the importance of having a webpage and social media outlets for your seed library. Otherwise, how will people

know you are there? As soon as you are online, you can join the Sister Seed Library list. Actually, you can be listed even if you have no web presence, but if people are trying to find you they will need contact information. If someone heard about seed libraries and wondered if there was one in their community, they could quickly find you on the list. If you are starting a seed library, taking a look at what others are doing is helpful. Please keep this in mind when you are setting up your Internet site(s). Knowing that others outside your community will be looking at what you are doing should encourage you to make your location clear. Include the name of your town or community and your state or province.

Your seed library website may be a page on the website for the public library, farmers market, or other organization. That might dictate what can be included and in what format, so make sure you are versed on any guidelines that need to be followed. I have looked at many seed library web pages. It is apparent that some are on government websites. I don't know if they are permitted to spice their pages up a bit with some graphics or photos, but it would certainly help. One of the most pleasant seed library websites I came across is for the Concord Seed Lending Library[1] located at the Concord Free Public Library in Concord, Massachusetts. This seed library was begun in 2013. Their logo is a seedling coming out of an open book. Nice!

A website should give information about your seed library and also give links to information from other sources. Many places are sharing information about saving seeds, and your website can both participate in the sharing and benefit from what others have shared. The Concord Seed Lending Library has pages labeled *home, hours and directions, resources, donors and supporters, events,* and *about us.* Look at many seed library websites and note what elements you would like to include in yours. Your website can be the base of operations for your seed library, with everything a patron needs to know to access seeds.

Putting your seed library on Facebook is something to consider. I found that, if they had only one, seed libraries were more likely to have a Facebook page, rather than a website. Many had both a Facebook page and a website. A Facebook page is simple to set up and there is no cost.

Put as much background information as you can on the *about* page on Facebook, including your location, with state or province. It should be a public page meant for an organization, rather than a personal page. Even people without a Facebook account can access a public page. People trying to find you shouldn't have to log into Facebook first.

The advantages of Facebook are that it is free, fast and easy to set up, and a message can easily be sent to followers. You need a Facebook account to become a follower, though. The disadvantages are that people get so many Facebook messages that, if they don't check them frequently, they can miss some messages. If that happens with your message, it is not easily retrieved unless the follower actually goes to the organization's Facebook page. Keep in mind that not everyone looks at Facebook.

Blogging platforms, such as WordPress and Blogger, are free. They could act like a website, but are used mainly for posting messages. The messages are archived for easy retrieval; either by month, subject, or popularity. I have a WordPress blog at www.HomeplaceEarth.wordpress.com. My posts are educational — mini-lectures, actually — and are always accompanied by photos to illustrate what I'm talking about. There are many ways people can receive my new blog posts. They can sign up to have it come by email so that whenever I send a post, it will come into their inbox. There are a number of other feed options that those of you more digitally enlightened than me would know about. A blog post won't get so lost in the digital mail as a Facebook post might. Every time I have a new blog post, I link it to a new message on my Homeplace Earth Facebook page.

A picture says a thousand words and brightens a page, making it more interesting. Use photos in all of these online venues. Having a place such as a website, blog, or Facebook page that needs pictures regularly will encourage you to document all aspects of your seed library in photos. That has been the case for me in my work documenting what I do in the garden. Note what you like about the photos on other websites/blogs/Facebook pages when deciding what to display on yours. Taking photos will help you personally focus on the subject at hand. If you use photos that someone else has taken, make sure you have permission to use them, and credit the photographer. If you use photos in which people can be identified, make sure you have their permission to use their likeness.

Graphics

A logo gives your project a recognizable identity. This is where the artists in your group will come in handy. You can use the logo when establishing your Internet presence and for brochures and signs. Take your time in starting your seed library, encouraging people to begin saving seeds of the open pollinated crops they are already growing, if the season permits. During that time of planning and deciding how best to proceed, the logo can be used on any notices or correspondence so it will be familiar when your seed library is ready to open.

The logo used by the Washington County Seed Savers Library is a colorful one — a basket tipping over with colorful vegetables spilling out — but it looks good in black and white, also. It is used on their brochure, signs, and website. The website was designed by the marketing and development officer for the library. The graphic work for posters and brochures was paid for by the Raymon Grace Foundation. Not every seed library has a marketing and development officer at their disposal, but artistic people are everywhere. Sometimes they just need some encouragement to show their talents. You could sponsor a contest to find an appropriate logo. This would be a good thing to do to help spread the word while you are still in the planning stages.

The artists you gather to design your logo will likely have ideas about other uses for their talents. A brochure printed on regular copier paper and folded in thirds is easier to take away than a full size sheet of paper. It should contain the information you most want your patrons to know. Identify your volunteers who have graphic design experience and ask them to help with this. If you don't have anyone with graphic design talents, put the word out to recruit someone. Some brochures include seed saving directions, such as the brochure for the Seed Library of the Pima County Public Library[2] in Tucson, Arizona.

Any graphics that are developed will be useful on your signs. You will need a sign to show patrons where the seeds are stored. It could be located on the seed cabinet or shelf, but a sign overhead can be seen from a distance to draw patrons to the location. You will also need notices to advertise the speakers and other events that will be involved with the seed library.

Education

If saving seeds were a common skill, you would not have to worry about providing education; but since that's not the case, your patrons will need help if they are to bring back viable seeds that are true to their variety. A public library has the benefit of being in the business of providing resource material. You will find a list of seed saving books in the "Resources" chapter at the end of this book. If your library doesn't already have books on seed saving, make a suggestion to the librarian in charge of acquisitions. Be on the lookout for new ones as they are published. A list of websites with seed saving information should be made available in the library and on any of your seed library websites or blogs.

Print and web resources are a great start, but nothing compares to hands-on experience. Plan to have presentations, demonstrations, and workshops about seed saving. Your educational initiative could begin with programs that bring patrons to the realization that saving seeds is an important thing to do. They can learn exactly how to do that later in the season. For Amanda West Montgomery (Pittsburgh Seed and Story Library), it was watching the movie *The Future of Food* that woke her up to the necessity of action. A showing of that movie or *SEED: The Untold Story* would be a good way to begin your awareness campaign. Another good one to show is *Open Sesame: The Story of Seeds*. These videos are listed in the "Resources" chapter. A video screening could be combined with follow-up discussions, demonstrations, or food tastings.

Identify the committee members in your group who are most suited to provide educational programs or to arrange for them. Make the best use of all the talents of those involved. If necessary, seek out more people with the talents you are in need of. A movie night could be a way to identify people interested in joining your effort to establish a seed library.

CHAPTER 7

Packaging, Signups, and Other Details

THERE ARE MANY THINGS TO CONSIDER when deciding exactly how your seeds will be distributed. If your intention is to educate and encourage seed saving, you will want to make that clear to your patrons. They should know it's not just about coming in and loading up on what they need for their garden for the season; the point isn't to just get for free the same amount of seed they would get in retail seed packets, but to create a sustaining seed saving community. It is like the Give a Man a Fish story. Give him a fish and tomorrow he'll be hungry and back for more. If, instead, you teach him to fish, he will never be hungry. Teach your patrons to save seeds, and they will be empowered to feed themselves and others. Your patrons may not be as successful in saving seeds as they anticipate, and that possibility should be expressed at the beginning. Let them know that even if they are not successful the first time around, they are welcome to come back next season and try again. It will take several years to develop a group of seed stewards — those who will dependably bring back seeds each year.

Packets or Bulk

Your seeds will be available to your patrons either packaged in packets (small envelopes) ready to grab-and-go or as self-serve. For the self-serve option, the seeds are in envelopes or jars and the patrons measure their allotted amount into small envelopes to take home. The grab-and-go option results in neat and clean displays and may be more attractive to patrons who have not had much experience handling seeds. It may also be more attractive to librarians who are still getting used to the notion of having seeds in their book library. However, it requires more work up front to pack the seeds from donations that have come in and to label all those packets. Prepacking seeds allows for barcoding each envelope, if that is your desire. More often, patrons record what they are taking or bringing back in a notebook, rather than checking their seeds out with the librarian.

The self-serve option requires the patrons to help themselves from the containers holding the donated seeds. They would need to be supplied

Bulk seed that patrons will put in envelopes themselves. Some bulk seed is in jars and some is in envelopes in the cabinet at the Goochland Community Seed Lending Library.

with envelopes, spoons for scooping seeds, a small tray to count seeds into or to catch any spilled seeds, labels, and a pen to fill in information on their packets about the seeds they are taking. Small envelopes can be found at office supply stores (coin envelopes) or sources of seed saving supplies — see "Resources."

How much you allow patrons to take will depend on how much seed you have available to share and what your objectives are. There are many ways to limit how much patrons take if you feel the need to do that. Some seed libraries limit the amount of seed or the number of packets a patron can take per visit or per season. Whether you have limits on how much seed each patron takes or not, your policy for that needs to be evaluated at the end of each year and revised as needed.

Labeling

Labeling is extremely important and should be present on all containers, large or small, that hold your seeds. Information to include on a label is:

* common name
* botanical (scientific) name
* variety
* crop family
* days to maturity
* the year the seed was packed for sale or grown locally
* when it was added to the library
* who it was acquired from and the location where it was grown
* germination rate, if known
* place for notes
* skill level

The seed packets, whether grab-and-go or self-serve, that the patrons take home should have this information, as well as the bulk containers. The packets that are pre-packaged can have computer generated labels — part of the extra work the volunteers would do ahead of time.

Seed envelope and rubber stamp.

Patrons helping themselves to seeds will have to label their packets themselves. A rubber stamp is helpful for this and can be used on adhesive labels or used right on the envelopes. Make sure the rubber stamp is sized correctly for the labels or envelopes you will be using. A stamp pad will need to be provided with the rubber stamp. Labels could be pre-printed with the form of what is needed, with the patrons filling in the necessary information when they add the labels to the envelopes. The rubber stamp used on the envelopes eliminates the need for extra labels. Check the stamp pad occasionally to make sure the ink has not dried out.

When patrons bring seeds back, the same information needs to accompany those seeds. The rubber stamp or fill-in-the-blank labels can be used for this new seed, with the person returning the seeds providing the information. Having seed catalogs available would help patrons and volunteers fill in the details. Some catalogs are not as detailed as others, so take time to determine if they will be helpful. If you have received donations from seed companies, make sure to have catalogs from those companies on hand. Common names can vary for the same crop.

Botanical names help to keep the distinction clear. They are particularly important to have with winter and summer squash. The squash genus *Cucurbita* has several species. Squash in the same species will cross pollinate. Most summer squash, zucchini, and some pumpkins are of the *Cucurbita pepo* species. You will also find *C. maxima, C. moschata,* and *C. mixta.* One of each species could be planted for seed saving and not cross pollinate. For example, one variety of each acorn squash (*C. pepo*), hubbard squash (*C. maxima*), butternut squash (*C. moschata*), and cushaw squash (*C. mixta*) could all be planted together and the seeds saved. Moschata squash varieties, by the way, are known to be a bit more resistant to squash bugs than the other squash species. There are many varieties of squash, and it is important to have the species name on the packet. Gardeners plagued with bean beetles on their common beans (*Phaseolus vulgaris*) will be interested to know that bean beetles are not attracted to cowpeas (*Vigna unguiculata*), also known as Southern peas. Each region has different insect pressure and weather conditions. Stocking the varieties that do the best in their unique area should be a goal of every seed library.

"Days to maturity" indicates how many days from the time of putting the seeds in the ground, or from transplanting, until you can expect to begin to harvest. Knowing the days to maturity is particularly important in garden planning if you are dealing with a short season or want to succession plant something in the same space after one crop is harvested. Of course, the days to maturity is only an estimate of when harvest can *begin.* It is up to the gardener to determine how long they will harvest a particular crop, which will dictate when they can plant their next crop.

At the beginning, the seeds in your project may come from seed companies. Once you start to acquire seeds from local growers, make sure to have the grower's name and general location noted on the seed labels. If a particular variety being offered is a collection from all the local growers, that information should be indicated on the packets or bulk containers. There may not be a known germination rate when a label is first made, but when a germination test is done, that information can be added to the label.

Different varieties of the same crop have different colors, uses, and flavors. Tomatoes, for example, might be red, yellow, or orange. The size can vary from cherry tomatoes for your salads to large slicers for your sandwiches. Some are meaty for drying and making sauce, and some tomatoes are juicy. Each has their own special taste. This is the kind of information that could be included in the space for notes on the label.

Skill Level

As with any new skill, saving seeds often involves some trial and error. Fluctuation of the weather will itself brings lots of challenges. However, the learning is in the doing. Encourage your patrons to put aside their hesitations and just go for it. Everything you can do to encourage them will be beneficial to all.

Some crops are easier than others to have success with when you are learning to save seeds. Providing this information for your patrons is a must. Some seed libraries use the categories *Easy, Medium,* and *Difficult* to distinguish the skill necessary to save seeds of each crop. Other seed libraries indicate the skill level involved with the terms *Super-Easy, Easy,* and *Difficult.* I prefer the word *advanced* over the word *difficult.* To me, advanced is more positive — if you save the seeds from the advanced crops, you have advanced your skill. The word *difficult* sends up a warning sign that may keep folks away. I'll use the categories *Easy, Medium,* and *Advanced* here.

Having too many choices can be overwhelming, and new seed savers may not know what they are getting themselves into. For the first year or two, you might offer only seeds in the Easy category, unless an experienced seed saver offers seeds in the Medium and Advanced categories. Fewer crops offered would allow the seed saving demonstrations that would happen later in the year to focus on just those crops. Also, having fewer crops, and emphasizing why that is, will call attention to the purpose of the seed library and that you expect the patrons to be saving seeds and returning them. The seed library at the Summers County Public Library in Hinton, West Virginia, bought their seeds to begin with, choosing varieties of the ten most popular backyard vegetables.

The varieties chosen were specific to what would grow well in the area. They added flowers later.

Easy. The crops that are easy for beginners to save seed from are those that have flowers with both female and male parts, allowing pollination within the same plant. You might find them referred to as *self-pollinators*. These are the plants in the Compositae or sunflower family (which includes sunflowers and lettuce), in the Leguminosae, or pulse family

Sunflowers are an easy crop.

(peas, beans, and peanuts), and the Solanaceae, or nightshade family (eggplants, peppers, and tomatoes). Patrons should be urged to begin their seed saving with these crops. Seed saving books will give minimum distances (isolation distance) to separate varieties of the same crop to prevent cross pollination. It should be noted that heirloom varieties of tomatoes (the ones with potato-type leaves) need more distance between varieties to prevent cross pollination than do modern varieties; this is due to the structure of their flowers.

Some of these easy crops are picked at the time the seeds are mature, such as tomatoes and sunflowers. Peas and beans, on the other hand, need to be harvested at the dried seed stage — when the seeds rattle in the dried pods. Lettuce needs significantly more time in the garden to grow much larger and taller before it sets seeds. Peppers need to ripen to their mature color; seed saved from green peppers will be immature and not germinate. This all needs to be accounted for in garden planning from the beginning, but it is easy to save the seeds.

Medium. The crop families that require a medium level of skill to save seed from are the Amaryllidaceae, or allium family (onions, leeks, garlic, and chives), the Chenopodiaceae, or goosefoot family (beets, chard, and spinach), and the Umbelliferae, or parsley family (carrot, celery, caraway, cilantro, dill, and parsley). These plants are *cross pollinators*. They are self-sterile and need other plants to cross with to produce seed. The wind and insects assist with pollination of these crops. You could grow different varieties in close proximity for eating without a problem if you only let one variety go to seed. Many, such as beets and carrots, are biennials needing to be in the garden for a second year before the plants produce flowers and set seed. Others, such as dill and cilantro set seed readily in the garden the year they are planted. (Cilantro, by the way, is the term for the *plant*. Coriander is the name for the *seeds* of the cilantro plants.) I like to overwinter celery and parsley in my garden. They pop back up early the next spring, producing flowers and then seed. Those spring flowers attract beneficial insects to my garden. In fact, the flowers of all the umbelliferous plants are good for attracting beneficial insects, which is an added bonus for your patrons growing these plants to seed so they can share them.

Carrot flower. Carrots are a "medium" crop.

Advanced. The crop families that require a more advanced level of skill include the Brassicaceae, or mustard family (broccoli, cabbage, collard, kale, and turnips); the Cucurbitaceae, or squash family (cucumbers, gourds, melons, pumpkins, and squash); and corn, which is in the Poaceae, or grass family. (Gramineae is the former name for this family.) Actually, saving the seeds from the medium and advanced crops is easy; taking the proper precautions to avoid cross pollination is the part that takes more expertise.

I enjoy growing both collards and kale for eating through the winter. Collards and kale will cross with each other, so in the spring, when they begin to bolt, I only let one variety of one crop flower and set seed. Since the seeds are viable for several years, I can save seed from one variety each year to add to my collection, giving me several varieties to plant for eating. Brassica crops are pollinated by insects, as are the Cucurbits. To ensure no cross pollination takes place in the squash family of plants, their flowers can be hand pollinated and taped shut.

Corn is wind pollinated, making the possibility of cross pollination from a neighbor a reality if precautions are not taken. It could be a big concern for those gardening in close proximity, particularly in a community garden. If you garden near large cornfields, you can prevent cross pollination between your corn and the corn your neighbor planted just down the road by using *time isolation* — timing your plantings so that the tassels from your corn are not shedding pollen at the same time as your neighbor's. Other techniques you can use are pollinating by hand or tenting your crop. Seed saving books will have information about all these techniques.

Wheat, barley, and sorghum are in the grass family, but they are self-pollinating, making them "easy" crops. These crops are not usually found in home gardens; however, with the growing interest in heritage

Bloody Butcher corn (red kernels) strung together waiting for shelling. Corn is an "advanced" crop.

wheat and in home brewing, they do have a place there. I grow wheat and sorghum in my garden as grains for eating and for compost materials. The wheat straw and sorghum stalks, as well as cornstalks, contribute carbon to my compost piles.

Making Seed Available

To hold the seed for your patrons, something as simple as shoe boxes can work, but most of the seed libraries I am familiar with have their seeds in cabinets with drawers. The seeds are either grab-and-go packets or self-serve. If they are self-serve, there will likely be more seeds in jars nearby. As I mentioned in Chapter 6, you could also keep your seed containers in boxes on a shelf. Wherever they are kept, they need to be organized. All the envelopes of the same crop should be stored together in the drawers and likewise, the larger containers holding the same crop should be together. You will have several varieties of each crop. The

Seed drawers labeled with skill level at the Carnegie Library of Pittsburgh-Lawrenceville.

crops in the same family should be together, since they will require the same skill level.

Labeling the drawers with a number, in addition to the crop family, helps to identify exactly what is in each one. Having the seed saving skill level on the drawers is helpful, also. There should be a chart nearby that shows the number of each drawer and what varieties you would find there. This chart, which would also indicate the skill level required, could be in the form of a poster or a laminated page. If the seeds are in drawers, you will need something to use as place-markers for your patrons to put in the drawers when they remove a seed envelope. A place holder should be large, and its purpose should be made clear so that it will be used. Otherwise, packets returned to the wrong place, or not returned to the cabinet at all, may be a problem. Nevertheless, someone

Place holder used at the Goochland Community Seed Lending Library.

should be assigned to go through all the seed storage areas occasionally to ensure that everything is where it should be.

If the seed is available as grab-and-go packets, other materials will not be necessary for the distribution. If patrons are required to fill the envelopes themselves, envelopes, spoons or small scoops, pens, and labels or a rubber stamp and ink pad need to be available. They could be stored in a drawer in the cabinet or in a decorative container on the table provided for working with the seeds. A small tray for counting seeds into and a container for seed packets that have been emptied assist with this activity.

Patron Notebook

If you were just distributing seeds and you didn't care how many packets went out and to whom, you wouldn't need to keep records. That

might be the case if you had an unlimited supply of seeds coming in. Even if you did have that unlimited supply, though, it would still be good to know which varieties are the most popular and which are not moving. Each packet could bear a barcode, as each book in the library does. That would require someone to record all those barcodes into an inventory. The advantage is that it is easy to keep track of your inventory with that method if all the packages are scanned when they leave the library.

The seeds in most seed libraries do not have barcodes. Their distribution is recorded in a notebook that the patrons maintain when they take out or bring in seeds. Each patron has a page with their name, their location, and what seeds they are either borrowing or returning. The information about the seeds being borrowed should include the date of the transaction, crop name, variety, year harvested or packed for sale, and source. The returned seed entry should include the date of the transaction, crop name, variety, year and location of harvest, and notes of any information that would be helpful to pass on.

To ensure privacy, the notebook that is available to the public should not contain private information, such as email and postal addresses and phone numbers. However, that information needs to be available somewhere. It could be stored separately by the seed library manager. That way, if the patrons need to be contacted for any reason, such as for upcoming events, it would be possible to do that. Occasionally someone might want to contact the grower of certain seeds. In that case, you would have to decide how the personal information of the grower would be made available. Public libraries already have the personal information of their patrons in their database, since it is required in order to obtain a library card. Check to see if you can legally access patron information through the library database. Not all seed libraries located in public libraries require their patrons to have a library card.

References to Have Available

The catalogs of your seed suppliers should be available for reference. A chart with gardening information, such as seed starting times, would be helpful. The Organic Seed Alliance, an organization that supports the growth and success of organic seed systems, has a wonderful 30-page guide — *A Seed Saving Guide for Gardeners and Farmers* — that is available as a free PDF.[1] It includes a very useful chart showing the common name, botanical name, and information about pollination, isolation distance, and population sizes. Other publications available from the Organic Seed Alliance include organic seed production guides for vegetables in the Pacific Northwest.

You will also find those Pacific Northwest guides, as well as guides for saving seeds in the Mid-Atlantic and South, at SavingOurSeeds.org.[2] Saving Our Seeds is a venue for raising awareness about our threatened genetic resources, and a resource for providing information and knowledge tools for seed saving and seed conservation. These publications can be copied and made available in a binder that is kept near the seeds.

Patrons will welcome a handout or brochure to take home that contains seed saving tips. Some of the seed libraries post their materials, including their brochures online. Helpful information can also be made available as posters or displays where the seeds are kept. Changing the display now and then will create interest and draw people to the location.

Make Your Own Seed Catalog

A printed list of all the seeds that are available and where to find them in your system should be available to your patrons near the seeds. You could even post that list on the website. Some seed libraries that have their seeds already packaged for grab-and-go allow their patrons to request what they want and pick their packets up at their local branch, even if it is different from where the seeds are stored. If that is how your seed library operates, you would want the list of resources posted online, as well as the seed list.

You could make a seed catalog specific to your seed library by cutting out the information for each of your varieties from regular seed catalogs and keeping it in a binder. Alternatively, you could print the page of the online catalog with that information to put in your binder. It might seem like an old-school way to do things, but it works. Encourage your seed savers to document their progress with photos during the growing year to make a catalog specific to your library. Those are the best photos to have for your catalog.

You can also use your photos on Pinterest, an Internet picture board. Some people maintain Pinterest sites on different subjects, posting pictures of things that interest them. I don't have a personal Pinterest board but have become acquainted with them when following a link from my blog back to where it came from. Besides posting only photos, you can post photos that link to other sites on the web. There are some really interesting Pinterest boards that people have put together — a feast for the eyes. You can create your own Pinterest board by going to www.Pinterest.com and using your Facebook or Twitter account, or your email, to set up your new Pinterest account. The Washington County Seed Savers Library uses Pinterest as their seed catalog, showcasing what is available. Photos of their activities can also be found there. When posting photos to your public sites, make sure you have permission to do so and always credit the photographer.

Learn from Others

Browse www.seedlibraries.net, and you will find labels, brochures, and signs to print out, along with helpful tips about organizing your seed project. That information has graciously been made available for your use through the work of Rebecca Newburn. At that website, you can view projects that other seed libraries have done and participate in a forum to address common questions and concerns. If you haven't already seen it, now would be a good time to watch the hour-long webinar *How to Start a Seed Library at Your Public Library*[3] from The Center for a New American Dream. The mission of The Center for a New American Dream is helping Americans to reduce and shift their consumption to improve quality of life, protect the environment, and promote social justice. Browse their website and you will find information to inspire you. In the seed library webinar, you will learn about the Basalt Seed Library in Colorado, the Pima County Seed Library in Arizona, and the Seed Library at La Crosse Public Library in Wisconsin. It's great to hear about these libraries from the very people who were responsible for getting them established.

Visit other seed libraries and talk with seed savers and gardeners. Make use of the Sister Seed Libraries list. If you can't physically visit other seed libraries, you can look at their websites to get ideas for your project. You will find seed savers by asking around in your community. If you are with a public library, a notice at the check-out desk or near the gardening books will do. Anyone who is connected with plants will be someone you can learn from. They may not have any more to share than the name of someone who would be interested in what you are doing, but that is something. Follow all leads to expand your network.

CHAPTER 8

Attracting Patrons

If you are already an established organization with a following, the simplest way to attract patrons to your seed library is to send out a notice through your regular channels, whether it be snail mail, email, posters, or word-of-mouth. If your seed library is not part of an established organization, you will need to do something to attract attention. Gardeners generally obtain their seeds in the winter when they are planning their garden. It allows time to start seeds of plants that will need to be transplanted. By the time Earth Day rolls around in late April, the time for planting the cool weather crops is past, but gardeners will still be looking for warm weather crops, such as beans, cucumbers, squash, and corn. You might keep that in mind if you have limited means to begin with, and you are opening after the planting season has begun. Lettuce can be planted as a fall crop for eating, but it wouldn't be going to seed in the fall. It bolts when the weather warms in early summer, producing seed in the hot weather. So, if you open in mid-summer and offer lettuce, don't expect to get lettuce seeds back that year. If you wait until Earth Day to open, you could distribute tomato and pepper plants, rather than tomato and pepper seeds. Otherwise, there wouldn't be enough time to grow out the seeds for transplant and a full harvest.

Start Early

Start getting the word out long before you are fully operational. You want to create a buzz in the months leading up to opening day. Many seed libraries spend six months to a year in the planning stage before they actually begin sharing seeds. Start talking about it while you are still planning. You don't have to have all the details yet, and you don't need seeds to turn people's thoughts toward seeds and seed saving. Keeping your project under wraps until you debut the seeds will only limit your project.

In Chapter 6, I gave several suggestions for fund raising. These projects, such as bake sales, yard sales, and silent auctions, provide start-up or continuation money and are newsworthy events to publicize. Also, they are a great way to meet people, some of whom will be potential volunteers.

Photography, Art, and Music

Some people garden all their lives and never save seeds. Many let their plants go to seed at the end of the season and never notice that it happens. If you want people to save seeds, they have to know what they are looking for. Sponsor a photography show of photos of plants gone to seed. Entries can be a series of photos of the plant at seedling stage, flowering, and at the seed stage. If the plants produce fruits and vegetables, a photo of those should be in the series. The seed stage may be the seeds on the plant (flowers that were never deadheaded) or the seeds in the fruit (a ripe tomato, cut open). In the case of cucumbers, the ones with ripe seeds are the ones left on the vine until they are big and yellow. Detailed photos of seeds with their pods and chaff, in the process of being threshed, are very instructive. Having photos will help the photographer/gardeners look more closely at what they are doing and serve to show others what is happening in the garden to produce seeds. Having an event to show off the photos will bring publicity to your seed project long before you even have seeds to share and will also serve to educate the public about where seeds come from. An event like this will help you identify seed savers who are also talented photographers.

Anything photographers capture can also be depicted by artists working in other mediums. The more creative people you involve, the better. You might even identify one or more talented folks to design a

Cowpeas growing next to corn. The cowpeas will need the strings to support the plants when they reach full maturity.

Mississippi Silver cowpeas. One pod and the dried seeds it contained.

logo, signs, brochures, and any other materials you will need to launch your project. Some of this launch material will have the potential to become posters to use in your seed library, so you may want to store them someplace after their initial use.

Sylva Sprouts Seed Lending Library

Jackson County Farmers Market
Sylva, North Carolina

Began in 2012

Mission: The Sylva Sprouts Seed Lending Library is a free seed project that has been established for Jackson County residents to come together as a community to share seeds, learn how to save seeds, and become aware of the importance of the seed and seed saving. By sharing seeds in our community we will develop seed stock that is well suited to our climate, mitigate our dependence on agribusiness and save money. By learning to save seeds we will foster our community's self-reliance.

This seed library is the project of Jenny McPherson, manager of the Jackson County Farmers Market. Each week at the market, she displays seed packets in a cardboard seed rack that formerly held seeds in a retail store. The bulk of the seeds and supplies are kept in a cabinet at the North Carolina Cooperative Extension Service Center where they can be accessed by patrons during weekday office hours. The cost to start was minimal, with seed companies, vendors, and local growers donating seeds to start.

Website: www.jacksoncountyfarmersmarket.org/sylva-sprouts-seed-lending-library-2

McPherson credits Rebecca Newburn and her outreach through Richmond Grows Seeds for the information and materials needed to make the Sylva Sprouts Seed Lending Library possible. Funding and encouragement came from the North Carolina Cooperative Extension Service Center, Appalachian Sustainable Agriculture Project, the Land Trust for the Little Tennessee, and the Jackson County Farmers Market.

Don't forget the musicians. Putting art, music, and gardening together really connects people. Along with the photography/art show, invite musicians to sing and play original songs about related topics. If they don't have anything original, you should be able to find songs relating to plants, gardening, and seeds to suggest. Now you really have an event going on. Add sidewalk chalk for kids young and old to decorate the concrete (with a plants/gardening/seeds theme, of course) and you are all set. If this event is indoors where chalk drawing on the floor is not an option, put large sheets of paper on the walls for the drawings. Add a plant sale and a bake sale to the day and you have a festival going on — and you don't even have seeds to share yet!

Join With Other Events

Having a table or booth at events such as local festivals or farmers markets will help publicize your seed library. Being at events was an important factor in getting the Carnegie Library in Pittsburgh and the Sylva, North Carolina, seed libraries established. In those cases, they had seeds to distribute to interested folks on the spot. Some seed libraries maintain a cabinet that is portable, in addition to one that stays put, just so they can do this. Although the Sylva Sprouts Seed Lending Library cabinet stays at the Cooperative Extension office and is available to patrons there, the seed library sees the most activity at the weekly farmers market. Prepared seed packets are displayed in a cardboard free-standing display rack that held retail seed packets in a big-box store. This display stand was destined to be discarded by the store at the end of the season, but found new service with the seed library. You will find a photo of it in the color pages in this book.

Gardening venues are not the only place to advertise your project. Health fairs are perfect for it. People know they need to eat more fresh fruits and vegetables and are thinking more and more about gardening. Free seeds would be attractive to them. As a way of advertising, you could hand out packets of seed — possibly something quick growing, such as sunflowers, zinnias, or beans — with no requirement to bring back seeds. Be sure to accompany the seeds with a brochure about your project. If the goal of your seed project is to get seeds back, make that clear. One disadvantage of being at events is that someone has to spend time and energy to be there. Time and energy are often the two things that are in short supply with busy volunteers or staff.

Educational information about your project, possibly using photos of plants gone to seed and of threshing methods, would be beneficial to have on display at these events; all the more reason to have a photo contest. Tri-fold displays are easy to transport and can fold up to fit in slim spaces when not in use. When I sold vegetables at the farmers market, I had a tri-fold display with photos of my garden. It attracted people enough to make them stop, look at the photos, and ask questions about what they were seeing. Often people don't know what to ask about without photos to prompt questions. Make sure each item or photo in your display is labeled. To keep my display from blowing over in the wind I cut about six inches along the folds to form flaps on the bottom. I folded those out and used C-clamps to hold the flaps to the table.

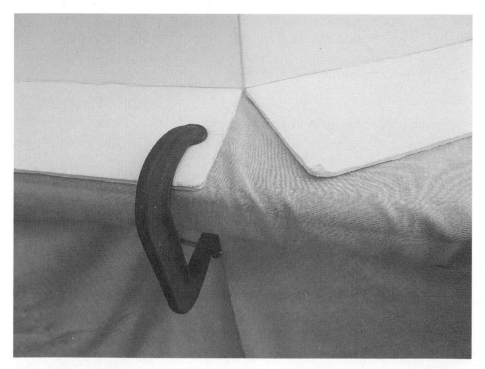

Clamp holding tri-fold display to a table.

Make Use of Volunteers

If you are one of those people who feel that you have to do everything yourself, you need to get over that. Not only is a seed library a big obligation, it is a project that benefits the community. Tap into the community as much as you can. I have already mentioned groups, such as Transition and Permaculture groups, that would make likely partners in a seed project. Even if they don't get involved in your project as a group, their members would be likely patrons for your seed library. Members of garden clubs, Master Gardeners, and participants in community gardens would also be interested. Have contact information for at least one person in each group. Keep them informed of your plans and ask them to spread the word.

Launch Party

When it is time to share your seeds with the public, have a party. Send out messages through your website, Facebook, and other social media. Call the newspaper, radio stations, and TV stations. In my research, I often found more information about seed libraries from online newspaper articles than from any other source. Make sure your opening is listed on community calendars. If connecting with the media is not one of your talents, find someone who is already familiar with the media through their job or other activities. Maybe they will volunteer to work on it, or at least coach someone on your team to do it.

Have refreshments, music, and a speaker or two. Have Fun! People might hear that there is a seed library starting, but will forget or procrastinate when it comes to getting involved. Having a set date with a celebration will get more people in one place at the same time. That allows everyone to meet each other. They will see people they already know and make new friends. If you are offering refreshments, which is always a good way to entice people to come to something, try to include some locally grown ingredients. The more you can do to promote locally grown food, the better.

Include a tour of the facilities, which might only be the seed cabinet, and include an orientation of how to use the seed library. If you have seeds from local growers already, a launch party is a good time to introduce those growers to new seed savers. Having the growers share their stories will provide a personal connection for your patrons from the very beginning.

Orientation

The Richmond Grows Seed Lending Library has an orientation video online, explaining how to use their seed library. If you have designed your seed library to work the same way, you might direct your patrons to that orientation, but it would be better to have one personalized to your project.

The Goochland Seed Lending Library began with the requirement that patrons attend an orientation in person to learn how to use the seed library. The session included tips about saving seeds. After the orientation, participants filled out a form to join, and only then could they borrow seeds. Orientation sessions were offered periodically. After the first year, that requirement was relaxed and the seed library became completely self-serve, with all the orientation material available in print with the seeds. This in-person orientation counted as continuing education credit for the county Master Gardeners. Since this seed library is located in the community college library, some, but not all, of the patrons are sustainable agriculture students who receive orientation instructions in their classes. In fact, some of the students may have participated in performing germination tests on the seeds.

An orientation would describe the program — how it came to be, the sponsors, etc. — and how it works. New patrons will learn how the seeds are categorized, how many they can take, and how to check them out. Everything will be clearly labeled and additional signs and posters can provide necessary information. This is a good time to talk about the details of seed saving so users have that goal clearly in their minds. Population sizes and isolation distances should be brought to their attention. Later in the season, you can have hands-on seed saving demonstrations, although having some at an orientation would be appropriate, also. Brochures with seed saving basics and helpful websites can be available to take home. Patrons should be made aware of all the book and video resources available in the library to help them have a successful year.

Online and self-serve orientations are convenient for patrons and for the staff. What is lost is the personal contact. Someone could borrow seeds, grow them out, and bring them back without ever talking to anyone. Personal interaction somewhere along the line is necessary to keep "community" in the process. After all, the overriding goal is to build resilient communities.

Continuing Education

Education needs to accompany any seed saving project. Offering free presentations and classes on seed saving and any other garden-related topic will attract new patrons. Keeping seeds in the hands of the people and out of the control of corporations is certainly a big part of why you should establish a seed library. In fact, for some patrons, that will be the main attraction. However, it is much more than that. It is about building a healthy community, from the soil on up. I love it when projects cross disciplines, and starting a seed library does just that. In the process of growing out the seeds they will receive, your patrons have the opportunity to learn about soil building, growing plants for healthy food, and growing plants for a healthy ecosystem. Programs on healthy eating, cooking, crafts from plant materials, local history connected with plants, and so much more can be geared toward attracting patrons to your seed project.

CHAPTER 9

Keeping the Momentum

Now what? That is the big question on everyone's mind once they have their seed library off the ground. Seeds have been borrowed, but none have been returned yet. If some have been returned, they have been trickling in with no promise of more for the season. The intent of a seed library is to have patrons grow the seeds out and bring some back. Without returned seeds, the program won't be able to continue unless there is another source of seeds. Seed companies have limits to what they will donate, so new seeds will need to be purchased, unless you can keep the momentum and excitement that started the program.

Patrons not bringing seeds back the first year is a universal concern for seed libraries. Even returns the second year are not guaranteed. If your seed library launches in a bad year for gardening, like so many did in 2013, there are even more challenges to overcome. To prepare your patrons to learn this new skill of seed saving and to be diligent enough to save seeds and bring them back is something that needs to be addressed from the beginning. As much work as you went through to get your seed library started with a location, seeds, and publicity, know that it is only the tip of the iceberg. Your seed savers need to be kept engaged and have opportunities for learning. When the launch party is announced, you should already have a list of presentations and classes planned for the year, or at least through the next few months. Something scheduled once a month is ideal. Even something scheduled quarterly would do. This chapter contains suggestions for events, but any gathering of seed saving patrons would suffice. A potluck dinner with no other agenda than for seed savers to meet and get to know each other would be good. The attendees can learn much from each other in a social setting and gain support just by knowing they are not out there alone.

Presentations and Classes

Presentations about all aspects of gardening and seed saving are important to educate your seed savers. Some may be new gardeners and need help with everything. The more experienced gardeners may offer to be the ones to give presentations or longer classes. I've already mentioned that Master Gardeners need to accrue volunteer hours. Those who are skilled in the subjects you desire presentations on may offer their services. The Master Gardeners also need to accrue hours of continuing education. They would welcome attending your classes.

The Washington County Seed Savers Library in Abingdon, Virginia, had already had presentations on organic gardening and seed starting and on the importance of seed saving for their patrons in 2014 before I visited in April that year to give a presentation about my new book *Grow a Sustainable Diet.* In May, they had regional expert Bill Best, author of *Saving Seeds, Preserving Taste: Heirloom Seed Savers in Appalachia,* on their calendar. Every time they have a program, new people show up in addition to the regulars, and when I was there it was no exception. They signed up some new seed saving patrons. Presentations are generally held in the large community room. The seed cabinet has its spot in the library for anyone to access, but it is on wheels, making it easy to roll to the community room to facilitate its use when the patrons are there. Circulation of all related material increased once the seed library was started at the Washington County Public Library. A library needs to stay relevant to its community. Once a program is established that brings people in, the library can justify adding more resource material and programs to serve those patrons.

Movie Nights and Book Clubs

Movies will attract an audience beyond your seed saving patrons. Everyone is affected by food, and there are a number of movies about the dangers of what is happening in the industrial food chain. Having a showing of one or more of these videos can move people to want to take action. Conveniently, you have a seed library there to fill their need. Building a new food system for the local community, which is what a seed library participates in, takes many hands. Most of us are oblivious to the big picture of what is going on around us. Once someone is made aware of the dangers lurking in the conventional food supply chain, they may feel helpless, until of course, they see the opportunities right in their own community — right in the library. Have a list of volunteer opportunities available for everyone, even those who are not gardeners. It is imperative that not all the movies depict doom and gloom. Make sure to show some uplifting ones that are about communities coming together to take control of their future. In Chapter 6, I mentioned some movies that would be good to start with. There are more listed in "Resources."

Most libraries already have book clubs that meet once a month. Everyone reads the same book and gathers to discuss it. At least that's the plan. My only experience with a book club was years ago when I happened to notice that a book club at the library in the next county was reading *Fast Food Nation* by Eric Schlosser. They were meeting the next week and anyone could attend. I was really interested in the book, but hadn't read it yet. All the copies were checked out, so I stopped at a bookstore and bought it on my way home. I wanted to be prepared and was excited to be among people who would be discussing it. Book club night arrived and about a dozen people came, apparently more than they usually have. I didn't know what they typically read, but quickly realized that it was not books like this. Most of those in attendance were rather detached from the overriding problems of fast food. Most of the regulars only read it because it was the book chosen for that month. One man who attended because he had small children and was concerned about their health came to the conclusion that irradiating food might be a good thing to rid it of anything harmful in the meat. With that in mind, I suggested that the food industry could irradiate manure and feed it to us. Another person there didn't seem alarmed by my suggestion and only asked if that would be nutritious. I mentioned some sources of local grass-fed meat, but the people in attendance didn't seem interested. I was hoping to find people with a passion for clean food, but only found people who were passive about what was happening. You can imagine my dismay.

My one-time experience with the book club was a good reminder for me about how other people think. My world is filled with people who are aware of what is happening with our food and want to make a difference. Unless people are alerted to things like this and have opportunities for action, they will continue to go through life with blinders on. There are many nonfiction books, like the ones Michael Pollan writes, that bring to light issues with our food system. They would be good book club selections. Will Stein of the Washington County Seed Savers Library was moved by Janisse Ray's book *The Seed Underground: A Growing Revolution to Save Seed,* which relates the stories of seed savers around the country and the heirloom seeds they keep in existence.

Not everyone is ready for such up-front information. They need to be eased into thinking about the food system as a part of their lives in a milder way. Fiction books that have organic farmers or others active in building a new food system as some of the characters in a story will help. Even a minor character like that could put the thought in someone's mind that there are such people out there. Any fiction books that have characters talking about seeds, participating in a seed library, or working in a garden are good ones to have on the reading list. Wouldn't it be great if the characters met up at the seed library, rather than the mall? Barbara Kingsolver includes garden-related ideas in her novels.

The book that put Kingsolver in the spotlight for championing local food is her nonfiction book *Animal, Vegetable, Miracle: A Year of Food Life.* It opened many people's eyes when it was published in 2007. At the same time, the book *Plenty: One Man, One Woman, and a Raucous Year of Eating Locally* by Alisa Smith and J. B. Mackinnon came out. Both books are about eating a local diet but approach the matter in different ways and in different locations. Smith and Mackinnon are from

British Columbia. In Canada, their book was published as *The 100-Mile Diet: A Year of Local Eating.* In 2014, Vicki Robin's book *Blessing the Hands that Feed Us: What Eating Closer to Home Can Teach Us About Food, Community, and Our Place on Earth* was released — more on local eating with yet a slightly different perspective. A book club interested in reading all of these books could have some good discussions.

I imagine there are book clubs with themes, and a book club could be formed with readers interested in gardening and sustainability topics. However, a book on one of those topics could be on the list of the not-topic-specific clubs. I recently noticed that a book club at my local library was reading *Black Potatoes,* the book I referenced in Chapter 2. The planning for movie nights and book clubs might function separately, but in cooperation, with the planning for the seed library. In other words, the same person isn't doing it all.

Stories and Histories

I am beginning to become aware of the number of people with an anthropology background who are working with seeds. Seeds and plants determine so much in our lives and the history of that is fascinating. We are wrapped in a web that encompasses everything. Rather than feeling trapped in such a web, we should embrace the connections we have and learn more about them. Providing books and videos about this is helpful. Having people come in to relate stories is even better. Someone might talk about working in the garden or hauling sweet potatoes to the market when they were growing up. Nothing can compare to hearing someone talk about their experiences firsthand. If someone were to talk about growing and marketing sweet potatoes at the library, having sweet potato slips to give to those who came to the talk would be a good thing to do. The talk could be planned at the time to plant the slips — once the soil is fully warmed. Sweet potato slips are the plants that grow from sweet potatoes if you just put the potato in a jar of water with half of the potato immersed. The potatoes could be started in water in the library about six weeks before the talk and slip give-away, allowing the patrons to watch the slips grow in the library. If the children's story hour included reading books about sweet potatoes during the time the sweet potatoes were producing slips, you would have done a good job of coordinating the library programs and attracting the parents and children to the seed library activities. Seed libraries can be about more than seeds.

Sweet potato with slips growing on it.

Photograph the seed savers who donate local seeds and record any stories they have to share, whether about current activities or stories about eating or growing this crop in years past. The seed savers who bring seeds to share can tell how they came to choose those seeds and their experiences growing and saving them. Even if they just received the seeds early in the year from the seed library, their story is still relevant. Not everyone has a friend or family member to acquire seeds from. Putting a face and a name to the seeds that will be passed on makes it more personable.

Children's Programs

Involving children in library programs, and hopefully their parents, is an ongoing task for librarians. Seed library promotion fits well with programming for children. Books about seeds, those who plant them, and gardens of all kinds can be read during story hour or put on the summer reading lists for the school-age children.

Seeds can be the theme of the summer programs planned for youth. If there are plants growing near the library that have gone to seed, take the children for a walk to examine them where they are growing and notice the seeds. Seeds could be harvested and brought back to the library for cleaning. Seed packets don't have to be limited to being rubber stamped with necessary information. The children can decorate one side of the packets themselves. A hand-drawn, hand-colored picture of a tomato, beet, or cucumber doesn't have to look like the exact variety to get the message across of what's in the envelope. Also, if the children have decorated the packets, the adults might be more encouraged to use them. When our youngest was in elementary school, the art department used that trick as a fundraiser to get the parents to buy magnets decorated with pictures their children had drawn. At the Fairfield Seed-to-Seed Library in Fairfield, Connecticut, preschoolers and autistic young adults decorated envelopes that would hold seeds to be checked out.[1]

Add history to the mix and have programs about Gregor Mendel, Luther Burbank, and George Washington Carver. Mendel, an Augustinian monk, proposed the principles of dominant and recessive traits in the mid-1860s as a result of his work with pea plants. Among the characteristics he studied were height of the plants, pod shape, seed shape (smooth or winkled), and the color of the flower. Pea plants could be grown in the library or in a garden near the library to illustrate what he was working with. The flowers of the peas Mendel was studying were either yellow or green. I have grown sugar snap peas with some varieties having white flowers and other varieties having purple flowers. Noticing these nuances would be a good activity for the children.

As for the height of peas, I enjoy planting Sugar Ann, a bush variety (short) of sugar snap peas, in the same bed as Sugar Snap, a pole variety (tall). The pole variety is planted along a trellis down the middle of a garden bed and the bush variety is planted on the outside edges of the bed. The bush variety ripens first and could provide a snack to eat while discussing all the characteristics the children notice between the two varieties at that stage. The pole variety will be starting to flower when the bush variety is ready to harvest. When the bush variety wanes, harvest can begin on the pole variety. Depending on the timing, this could be done with children in a summer program if it was done before the weather becomes too hot for peas. The days to maturity of the chosen varieties would have to be looked at closely and planned accordingly. Beans could be substituted as a summer crop in hot weather. More than one variety of each bush and pole pea/bean variety could be planted for more comparisons. Having something like this can bring a book to life. The observers can engage all their senses. Mendel's work was not recognized until after his death, which is another good lesson to pass on to children. The work of many people who have contributed to our society, whether they are scientists, artists, or anywhere in between, was not recognized until after their death. Rather than seeking fame and fortune, they were working from their heart.

Luther Burbank was a self-taught plant breeder. In 1930, four years after his death, the Plant Patent Act was enacted and 16 patents were awarded to him posthumously for plants that he had developed. His work, plus his book, *How Plants Are Trained to Work for Man,* served as an inspiration for the Plant Patent Act. That information might be a bit much for the children, but could spark some good discussion among adults.

George Washington Carver's history is fascinating. Born in 1864, he rose to become one of the most prominent Americans of his time before he died in 1943. He was highly educated, a rarity for African Americans at the time. Carver's work with peanuts and other crops at the Tuskegee Institute, helped many struggling share-croppers throughout the South. Your library activities could include peanut plants for observation and for tasting. Peanuts grow underground from "pegs" that the plants put out after flowering. The products that George Washington Carver developed from peanuts could also be studied.

Work Parties

Sometimes you need to gather people to take care of the chores. Instead of mentioning work, you can call the gathering a party; such as a seed cleaning party or a seed packing party. Seed cleaning is a skill to be learned by new seed savers. The work party can be instructional and can be a time and place for them to bring their own seeds to be cleaned — threshed from their pods and winnowed in front of a fan. When I save seeds, I use a variety of things that I already have; such as colanders and strainers with holes of different sizes. You can buy special screens to facilitate cleaning seeds and, if you have the funds, you might choose to do that. You will find photos of seed screens on the color pages in this book. Whether you have specialized equipment or improvised items, make the most of the activity. Packing seeds in envelopes from larger containers would teach your volunteers the reasons behind how many are put in each envelope. Seeing how the seeds are brought back can provide lessons in how to do it better the next time.

If you have only a small amount of seeds to clean or to pack, it is sometimes easier to do it yourself, but remember that this is a teaching opportunity. The more your patrons understand about the process, the more skilled they will become. Even if only a few people show up, you are building community.

Eat the Bounty

If your patrons are growing food crops for seeds, they will be producing food. In most cases, well before there is a seed crop, there will be food to eat. In the case of tomatoes and peppers, the seeds will be ripe when the vegetables are ripe. Hopefully, enough will have been planted that some of the food can be shared with the community. You could have a soup day where everyone brought something from their garden appropriate to add to the soup pot, whether it was grown from seeds acquired from the seed library or not. If the seed library maintains a garden, contributions can be made from that. This soup could be made for a meeting of the seed patrons or it could be available to a wider audience of eaters to publicize the seed library, depending on how much soup you have (and how many hungry patrons).

Potluck dinners are in order whenever you have a meeting. I like the ones where everyone brings their own table service of non-disposable cup, plate, silverware and napkin, in addition to their food to share. A community of seed savers should be able to have a table full of dishes that contain something grown in the gardens of those supplying the food. Not everyone is a prolific grower, so even if it is a little basil or mint from the garden, or edible flowers added to a dish, it would suffice. You could go a step further with the soup or potluck and have small signs about the varieties of each crop that was contributed, particularly if it was an heirloom. Homegrown flowers can grace the table.

Promote Seed Gardens

Seed gardens are gardens where the main crop is seeds. When people talk about their gardens and their harvest, rarely do they talk of seeds as a harvest. Usually, it is the vegetables, fruits, and flowers that they are talking about. You will have a hard time getting seeds back if seeds are not considered the harvest from the get-go. The whole garden doesn't have to be designated to save seeds from, as long as seeds are planned for and saved from some part of the garden.

Having plants in the landscape around the library providing seed for the seed library is ideal and will serve as a public visual aid for the library. You would want to put a sign up to designate those plants as being grown for seed so that an overzealous volunteer doesn't take off the spent flowers or pull out the plants thinking they are doing you a favor. Beauty is in the eye of the beholder. Patrons need to be educated about the beauty of a plant growing all the way to seed. Having the collection of photographs that I suggested in Chapter 8 would assist in doing that.

Connect with school and community gardens in the area. A special orientation for those planning or using school and community gardens might be in order. A representative from the seed library could help

Lettuce going to seed. Beneath each white puff is a group of seeds.

them plan what seeds will be appropriate for them to "borrow" that they could let grow to seed in the time and space they have allotted. School and community gardens each have their special considerations. Unless school is in session year round, there is a more restricted time-frame that needs to be attended to. From my experience volunteering with the garden program at my children's elementary school, teachers struggle enough with managing a garden that has food and flowers when they need them, let alone planning for seeds. There have been many good changes in the area of school gardens in the two decades since I was involved, but I imagine that planning for seeds is still a challenge. School gardens are wonderful additions to the school curriculum. Children can learn much more, and retain it, if they learn with all their senses. Smelling, feeling, and dissecting a flower has much more meaning than learning the parts of a flower from a worksheet. Knowing that not everything can be grown at the same time is enlightening for many. Tasting sugar snap peas freshly plucked from the vine and pulling a carrot out of the ground are among the experiences everyone should have at an early age. Learning the full cycle of those plants is a more holistic education, rather than a fragmented one of seeing them at only one stage of their life.

Most restricting for community gardens is space, rather than time; also the gardeners themselves. Gardeners can be pretty independent individuals, even if they are all members of the same group — in this case, the community garden. If the garden manager decides that it would be good to have one bed dedicated to growing seeds for the seed library, the manager needs to be sure those seeds won't cross pollinate with plants the gardeners are growing in close proximity. The gardeners themselves would have the same problem if they wanted to save seeds in their own plots for themselves or for the seed library. Isolation distances can't be met in such small spaces, so other measures need to be taken. Isolation by time is effective, but that would take much planning and cooperation among all the gardeners. I've used time isolation to grow more than one variety of corn without having it cross by choosing varieties that have different maturity dates and planting them at the same time, or staggering the planting dates by at least two weeks if the maturity dates were the same. Netting will work with some crops to keep the insects from cross pollinating your crops. In some cases screened cages could be built. As I mentioned in Chapter 7, hand pollination, particularly with squashes, is done to prevent cross pollination, with the flowers taped shut afterward. You'll find all these techniques in the books on seed saving.

Seed Stewards

Encourage the development of a Seed Steward program at your seed library. Seed stewards are individuals experienced in seed saving who take it upon themselves to be guardians of certain varieties. They can be counted upon to return a quantity of seed well above what is required by the regular patrons. This is particularly helpful if you are trying to maintain certain heirloom varieties unique to your area, but you can have seed stewards for any crop. Having seed stewards will relieve the pressure of wondering if you will get seeds back from all the patrons. The extra seeds supplied by the stewards can also be ones that you give away freely, if that is something you feel called to do. Especially in the case of heirlooms, this could be a way to have them grown in the school and community gardens without the requirement to save the seeds.

Seed stewards will be your most reliable growers. They could be your most reliable volunteers, but not necessarily. I know some long-time seed growers who shy away from being involved with groups, but are more than willing to share their seeds. Recognition of your seed stewards could make that position something that new seed savers aspire to.

Pace Yourself

Beware of volunteer burnout with yourself and with others. When people jump into a project wholeheartedly they may reach a point where, as much as they would like to, they can't do it anymore. Rob Hopkins (*The Transition Handbook*) mentioned the potential for that very real problem in his talk at the Whole Earth Summit in March 2014. Try to spread the responsibility for different parts of your project to others.

For me, a successful project is one that I may have worked diligently to get established, but ultimately can operate without me.

After the second year, you may find that you have seeds that were donated the first year that were never used. If you do a germination test, it would probably show less than desirable results. You need to purge those seeds from your offerings. If you feel that they could still do some good for someone, offer them in a bin or a box as free for the taking — free

Washington County Seed Savers Library

Washington County Public Library
Abingdon, VA

Began in 2013

Mission: A seed library will help to build biodiversity for our local food supply. As part of this effort the library will offer educational opportunities on the importance of a viable local food supply, the benefits of using heirloom (vs. hybrid) seeds, seed starting, seed saving, gardening for beginners, gardening in a small space, and composting.

This seed library was begun by Will Stein, reference librarian at the Washington County Public Library, with the help of key advisors Ben Casteel of Appalachian WildSide and Heather Jeffreys of Appalachian Sustainable Development. Casteel and Jeffreys offered classes the first year. Classes for patrons in a variety of gardening topics continue at the library.

A brochure with seed saving tips is distributed to seed patrons. Seeds are in packets for patrons to take. Seed packets are stored in an old card catalog that was painted by a volunteer. This cabinet is on wheels, making it convenient to move to the community room when programs (and seed patrons) are there.

Seeds were purchased to begin the seed library. Funding for the seed library is provided by the Raymon Grace Foundation.

Website: www.wcpl.net/seed-savers-library.html.

Pinterest is used as a "seed catalog," with photos posted of the varieties of vegetables that are available as seed from the seed library.

meaning free of obligation of bringing any seeds back in return. Make sure it is understood why they are there. If the germination is too low, don't even offer them as a freebie.

Although seeds may be slow coming back the first year, the returns will pick up if you are diligent about keeping in touch with your seed savers and making them feel part of the process. Having events, such as the ones suggested in this chapter will do that. These ideas are given with public libraries in mind. If your seed library is located in a business, museum, or other venue, you should be able to adapt these activities to your situation. I hope they will inspire you to think of new ideas that will be exactly what you need.

CHAPTER 10

Seed Swaps and Other Means of Sharing

T HE MOBILITY OF PEOPLE is such that they may plant a garden but are not still around to collect the seeds. With others, saving seeds to bring back to a seed library seems like a responsibility they are hesitant to take on. There are other ways of getting seeds into the hands of the people, such as a seed swap, in all its various forms.

Community seed saving and sharing initiatives don't have to be as organized or involved as a seed library. It could start as a group of friends trading seeds on a winter afternoon. Sometimes friends come together to make a seed order, either to order in quantity to get bulk prices or to decide on packet varieties to share, since there are often more seeds in a packet than a single gardener can use. Seeds left from the previous year can also be shared. Once the members of the group begin saving seeds, those can be added to the mix.

This type of activity is generally known as a seed swap. If you are participating in a seed swap, you only have to save seeds from one or more varieties, not from your whole garden. You will have more seeds than you need of those varieties you saved, so you will have some to share. Some people specialize in saving seeds from certain plants or plant families. When everyone gets together, there are varieties of seeds enough for a diverse garden — as long as enough gardeners are saving enough seeds. If a swap will be your only source of seeds, it takes trust in the other members of your seed sharing group that they will come through with the seeds you need. In turn, they will be trusting that you will do the same. You may want to be a member of the group in order to be introduced to varieties new to you, not for seeds for your whole garden. A seed library is actually a version of a seed swap.

Expand that activity beyond your group of friends a bit, and you could have members of an organization come together to make a seed order (or not), trade leftover seeds, and share saved seeds. The next step is to open it up to the community. Once you get to the community level, you will probably need a larger meeting space, more leadership, and publicity. If you intend this to be more than a one-time event, you will want someone to document the occasion so it can be replicated. When it comes to seed swaps, I prefer to use the word *share,* rather than *swap.* When people hear the word swap, they think they need to bring seeds to trade to get seeds. Although that might be how it works with some seed swaps, it is not always the case. The word *exchange* also implies that you need to bring something to trade. Swaps are actually seed sharing events. The seed swaps I've attended offer seeds to anyone who comes, with or without seeds to share.

One of the cowpea varieties I grow came from a seed swap/share. I tend to call that variety of cowpea Arkansas Razorback, although I believe the name attached to the seeds was actually Ozark Razorback. Our daughter lived in Arkansas for several years, so I guess I quickly forgot Ozark in favor of Arkansas. Seed names are a tricky thing. I can see how easily they can get changed, just as I did with the cowpeas. Some original names are forgotten over the years, and when the seed was documented, it is often named after the seed saver who kept the line going. Seed Savers Exchange deals with many varieties of seed received from different sources that could possibly be the same. You might run into that sort of thing when others are sharing seeds with you.

Southern Exposure Seed Exchange (SESE) is a seed company in Mineral, Virginia. They often host a seed swap at conferences and events they attend. At these sponsored seed swaps, I have received

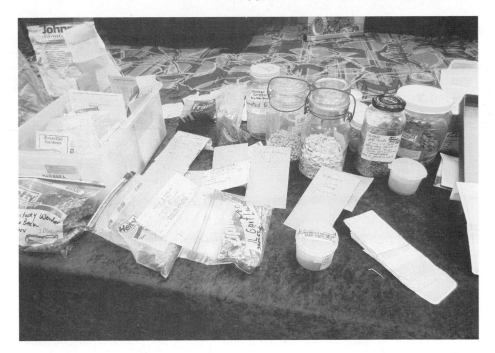

Seeds available for self-serve at the Virginia Biological Farming Conference seed swap.

some great seeds: a new (to me) variety of peanut, and the Arkansas (Ozark) Razorback cowpeas I mentioned earlier. SESE and individuals with seeds to share put their offerings on the large table. Guidelines are posted and envelopes are provided for those acquiring the seeds — it is a self-serve operation. The space and publicity are already provided by the conference. As far as I know, there is no documentation about what is donated or dispersed at these events. In those settings, it isn't necessary to collect names or record varieties. In other situations, it may be an advantage to do so. A seed swap/share could be just the thing to help stock your seed library. The Portland Seed Library[1] in Portland, Oregon, has an annual seed swap where donations are accepted to stock the seed library.

Seedy Days

In my research for this book I kept coming across references for Seedy Saturdays and Seedy Sundays — seed exchanges where sharing seeds was the main event. Seedy Saturdays were usually in Canada, and Seedy Sundays were usually in the UK, but not necessarily. You could declare any day a Seedy Day and host a seed share. These Seedy Saturdays/Sundays are more than just a table to browse and take what you want. The growers are on hand to give advice and there are speakers and vendors. The whole day is a celebration of seeds. The Seeds of Diversity website maintains a list of Seedy Days in Canada[2] and offers information to help you organize one of your own. These seed share days are characterized by low cost, local talent, and volunteer energy.

The first Seedy Saturday was held in Canada in 1990, the brainchild of Sharon Rempel. Roy Forster, curator of the VanDusen Botanical Garden in Vancouver, Cathrine Gabriel from Health Action Network, and Dan Jason from Salt Spring Seeds helped her create the first event. In addition to seed savers exchanging seeds, small-scale seed companies sold open pollinated seed. Rempel invited agronomists from the University of British Columbia to the event. Her hope was to get the heritage varieties of seeds kept by backyard gardeners trialed and evaluated regionally, and a core collection of regionally adapted vegetables, fruits, and grains conserved and exchanged annually. In addition to the Swap Table at the event, a diversity of talks informed the public about the international politics surrounding seed control.

Rempel hoped that Canadians would become involved in setting up community seed banks that stored quantities of open pollinated seeds. She also hoped people would share the cultural legacy of their seed, and that this traditional knowledge would be conserved with the seed in community seed banks.

When Rempel started Seedy Saturday, the Canadian gene bank (Plant Gene Resources Canada, or PGRC) had one variety of lettuce in its collection. The urgency to get more diversity in the bank became Rempel's career path. She wanted the seed that people had brought with them from the "old country," varieties that were part of cultural heritage globally. The food, the seed, and the story are all interconnected. It was through her work that Red Fife wheat was brought back into production in Canada. It had been the wheat that fed Canadians from 1860–1900.

Good ideas spread like pollen in the wind. Through the collected talents of Rempel and her friends, a seed sharing event was begun in Canada that eventually extended to the UK. Seedy Sundays in Brighton, England, began in 2002 after two members of the Brighton and Hove Organic Gardening Group vacationed in Vancouver and encountered a seed swap. They took the idea home with them. You can find a list of other Seedy Saturday/Sunday events in the UK, as well as suggestions for organizing your own, at www.seedysunday.org.uk.[3]

National Seed Swap Day

In the US, the last Saturday in January has been declared National Seed Swap Day, thanks to the efforts of Kathy Jentz, editor/publisher of *Washington Gardener* magazine. The first annual *Washington Gardener* Magazine Seed Exchange was held in Washington, DC, in 2006.[4] Making it an official day spreads the idea faster and encourages others to join the fun. There are seed share events tagging on to National Seed Swap Day around the country. You can find where they are at www. seedswapday.com. If you can't have a seed swap on the last Saturday in January, have it a different day. You could even come up with a different name unique to your event. A seasonal or year-round seed library could grow from an annual seed swap/share day once the public interest is there and a core of seed savers has been identified.

Seed Banks

When I told people I was working on a book about seed libraries, they would often begin talking about the seed bank in Norway.[5] Officially known as the Svalbard Global Seed Vault, it is often referred to as the Doomsday Vault. Seed companies from around the world have seeds stored there which are duplicates of seeds they have stored elsewhere. I do not feel comforted by the fact that if worse comes to worst, we can still retrieve seeds from an island in Norway to continue civilization as we know it. As you well know, seeds are living things that need to be grown out periodically to stay viable. Granted, the experts have taken great care to make sure the seeds in Norway are stored very dry and very frozen so they will last a long time, and I imagine they will be replaced with new samples if needed, but I still am not reassured. Keeping the seeds in production and keeping the supply in many hands, jars, freezers, and gardens is what reassures me.

Some people are concerned enough about their food supply to start their own local seed bank. They identify what their community would need to survive on and begin to store seed samples of those crops, choosing the varieties that would do the best. If something happened to disrupt the food supply, a community would need to rely on only what they could grow. This is different from a seed library, where the goal is to get seeds out to as many people as possible. Our communities need both initiatives, and I can foresee that the lines between the two will begin to blur in the future as seed libraries gradually develop some aspects of seed banks.

Dan Jason at Salt Spring Seeds on Salt Spring Island in Canada has long been concerned with the ownership of seeds, biodiversity, and communities being able to feed themselves. In 2002, he and some friends started Salt Spring Seed and Plant Sanctuary,[6] a nonprofit foundation that will be "a living gene bank for all the edible and medicinal plants that can be grown in Canada."[7] The Sanctuary maintains a database for all the seeds and plants in its collection. The name "sanctuary" was chosen over "bank" because seeds are holier than money. I like that.

The first order of a seed bank is to gather as many varieties as it can and to keep the seeds viable by occasionally growing them out. Now that they have the varieties they need, Salt Spring Seed and Plant Sanctuary is embarking on the next step, which is to build up their supply so that the community would have enough seeds to actually grow out to feed itself. Keeping samples of everything holds open the possibility for feeding a community. Having enough quantity of everything yields the promise of that happening.

Some of the members of the Comox Valley Growers and Seed Savers[8] (CVGSS) on Vancouver Island in Canada started a seed bank with encouragement from Dan Jason and funding from the CVGSS. They began with the members first growing out seeds that they supplied themselves, in contrast to a seed library starting with seeds to give away. The first returns came in the fall of 2012 and were stored in containers in two plastic totes, each holding an identical seed set. Each tote was stored at room temperature in a different place. Every seed contribution is accompanied by a Crop Record form. A binder with the Crop Record forms for all the seed was packed in each tote. One set has since been moved to a freezer to ensure safekeeping, since it is hard to keep everything grown out as often as they would like.

The Comox Valley Seed Bank has established a Seed Curator program for individuals to take on the responsibility to grow out certain varieties, much like the Seed Stewards I've suggested. The Seed Curator signs an agreement to use good seed saving practices, avoid synthetic fertilizers and pesticides, and collect and record the data needed for the Seed Bank Crop Record form. These seed banks I've mentioned are "living" seed banks, with care taken to grow out the seeds regularly.

COMOX VALLEY SEED BANK CROP RECORD

Common Name (e.g. Bean)	Grower code:
Botanical Name	
Variety (e.g. Black Turtle Bush)	Year collected:

Seed source

Seed company	Other Source
From Seed Bank Stock - Year Collected _____	Seed Bank Grower_____

Growing Information

Date planted in flats Number of days to germination	Number of plants grown
Date planted out	Outdoors / greenhouse / cloche
Approx germination rate (%)	Productivity
Date ready to eat (even small)	Number of days seed to food
Date seed collection began	Number of days seed to seed

Resilience to extremes (e.g. stunted, slowed, rotted, wilted, no problems) Include relevant weather information

Diseases/pests	
Heat/cold	
Rain/drought	
Other relevant information (flavour, ease of threshing, etc)	
Measures you took to prevent crossing with other members of the same family	

Seed harvest

Amount donated to Seed Bank (cups/spoonfuls)
Storage conditions for seeds you kept at home

The Seed Bank is a project of Comox Valley Growers and Seed Savers. Its purpose is to preserve and maintain a collection of viable, open-pollinated, non-GMO, organically grown food seeds that are well-adapted to the growing conditions of the Comox Valley.

Comox Valley Seed Bank Crop Record form.

It wouldn't do anyone any good to have a large stock of dead seeds. These projects are only a start, and I'm sure it has taken much longer than Sharon Rempel planned when she envisioned seed banks all over Canada, but you have to start somewhere.

When you live on an island, such as Salt Spring Island or Vancouver Island, and supplies come in on the ferry, the subject of food security takes on a special meaning. If you think of it, every place could be identified as an island of some sort, in a broader sense of the word. Cities that depend on food being trucked in are at the mercy of the major roads staying open. Even cities that are not totally surrounded by water might have bridges as major access routes. When traffic has to cross a bridge to reach a city, the future of regular supplies coming in is even more precarious in these changing times. There is talk of food deserts in the midst of cities where there are no stores that carry fresh food, only packaged items. We need to have gardens, and the seeds to plant them, everywhere, in order to feed the population.

In my book *Grow a Sustainable Diet,* I help you begin to plan what it would take to grow your food, including the cover crops needed to feed back the soil. The worksheets I provide help you calculate how much seed you would need for each crop. You can take that one step further to decide how much extra space would be needed to grow the seed for that much food. The book *Hungry Planet: What the World Eats,* by Peter Menzel and Faith D'Aluisio, shows families from around the world in their homes with the food that is typical of a week's worth of meals for them. These authors also wrote *What I Eat: Around the World in 80 Diets,* showing individuals and what they eat in a day. What would your table look like if you laid out all you eat in a typical day or week? If you were to plan to store seeds that would be capable of producing the food for your community for a year, there would need to be much discussion and soul searching about what food and how much of it would be needed. You never want to plant all your seeds at one time in case something happens to that planting or to the harvest, so you would need even more seeds as backup. You never want to put all your eggs in one basket, lest something happens to damage them all at one time; trusting nations or corporations to be the custodians of our seed is the same thing. Small seed banks in every community, with many seed savers regularly growing out the seeds to keep a fresh supply will result in resilient communities ready for whatever the future holds.

I mentioned Navdanya and its founder Dr. Vandana Shiva in Chapter 1. The setting up of community seed banks is central to Navdanya's mission of regenerating nature's and people's wealth.[9] By 2014, Navdanya was responsible for 120 Community Seed Banks (CSB) being established in 17 states across India. The seeds in each CSB are unique to the community they serve and essential to helping the small organic farms increase their biodiversity and their productivity. Educating and empowering the people is what Dr. Shiva is all about. By conserving the seed varieties grown in India before the global corporations took over, Navdanya was able to distribute salt-tolerant seeds in Orissa after a super cyclone hit, facilitating the rejuvenation of agricultural biodiversity there.

When Monsanto came to India, GMO seed was given to the farmers in exchange for their open pollinated seeds. When the GMO cotton crops failed, Dr. Shiva and Navdanya were able to give farmers open pollinated seeds to grow once again from seedstock they had kept in production. Without Navdanya, the farmers would have had nothing to fall back on. Education about saving seed is given along with the distribution of seeds. If your enthusiasm for seed saving and working to keep seeds in the hands of the people begins to falter, take a few moments to check on what Dr. Shiva is doing.[10] You will come to realize that this may be one of the most important things you will ever be involved with.

Share What You Have

If you have access to lots of seeds and just want to share them with others, you can do that. The group Eating in Public in Hawai'i does that with Seed Share stations that they leave wherever people will have them — community centers, libraries, churches, coffee shops, senior service centers, etc. The recipients of the stations are asked to commit to looking after them for at least a year. Eating in Public believes that saving and sharing seeds is crucial to our freedom, autonomy

Portable Seed Share station made using the plan from Eating in Public as a guide.

Portable Seed Share station with sliding lid used as a backdrop for potential signage.

from capitalism, and for our survival. The Seed Share stations are built using scrap and repurposed material and are launched with a starter kit of recycled envelopes and approximately 50 seed packets. You can build your own Seed Share stations from the plans they provide on the Internet.[11] The plans are available in English, French, and German. The Seed Share stations freely offer seeds to anyone, without an obligation to bring any back. However, in the spirit of how Eating in Public works, with people caring for people (and the planet), seeds will likely be deposited at these stations, not just taken. Some seed libraries would like to have a portable component that would enable taking seeds to public events. The plans for the Seed Share stations would work well for that purpose.

Seed Share stations remind me of the Little Free Libraries that have sprung up in many communities. A Little Free Library is a sheltered place with books free for the taking. It might be a box on a post near someone's home, near a community center, or anywhere. People can take what they want to read and leave what they want to share. Most likely, more people have books to share than seed they've saved, but that could change.

Start a Seed Company

You can grow seeds out for your own use and whatever happens is okay. When you begin to grow seeds for others to contribute to a seed library, seed swap/share, or a seed bank, you must accept more responsibility to make sure those seeds are what you say they are. You need to document what you have done and make sure they haven't cross pollinated with anything, unless of course, you intended that hybridization, which would be noted. Education is necessary in all aspects of this process. The more you know, the more you realize there is still more to know. It can be quite an exciting journey. You can get so involved with seeds that you begin to grow a few varieties for a seed company. In *Sustainable Market Farming*,[12] Pam Dawling includes a chapter on seed growing. The next chapter, "The Business of Seed Crops," is written by Ira Wallace[13] of Southern Exposure Seed Exchange and explains how to go about growing on contract for a seed company.

If you excel at growing and marketing seeds, you may decide to start your own seed company. In the Winter 2014 issue of *Seeds of Diversity* member magazine, Bob Wildfong reported that many of the Seeds of Diversity members have started their own small heritage seed companies as a result of their seed saving hobby, no doubt fueled by Seedy Saturdays.[14] Canada has about twice as many small-scale seed companies per proportion of population as the US has.

Caitlin Moore and Claire Ethier (Olympia Seed Exchange, mentioned in Chapter 1) have started the Root and Radicle Seed Company, growing out their seedstock in 2014 in preparation for their 2015 opening. Their seeds will include vegetables, flowers, and herbs adapted to the maritime climate and low-input organic systems.[15]

You may recall that I mentioned the Hudson Valley Seed Library in Chapter 1. The Hudson Valley Seed Library had its beginnings as a seed library in a public library — the first public library to offer seeds. Things change, and it is now a seed company with some interesting aspects. You don't have to be a member to buy seeds from them, but if you do sign on as a member, you receive a discount on your purchase and a packet of seeds that is part of their Community Seeds: One Seed, Many Gardens program. Through that program the members receive seed saving instructions for the seeds for the current year's project. All the seeds that are returned at the end of the season are pooled together and donated to garden projects, such as school gardens, food banks, and community garden groups. Members can nominate who they would like to receive the free seeds. At this writing, there are over 1,200 members from all over the Northeast and beyond!

Another unusual aspect of this seed company is its art packs. Some of their seeds come in packets that have designs that are actually commissioned art. Each design is by a different artist, and new designs are added each year. Most of their artists are gardeners. The Hudson Valley Seed Library specializes in seeds that do well in the northeast region of the US They are doing as much of the growing out and seed breeding as they can at their small certified organic farm and rely on local farmers, farmers in other regions, and trustworthy wholesale seed houses that are not owned by or affiliated with multinational biotech companies for the rest.

Seed companies are in the public spotlight and can help publicize the plight of seeds. The varieties they choose to sell will be the varieties the average person has available to them, unless they are participating in seed libraries or such organizations as Seed Savers Exchange and Seeds of Diversity Canada. The more publicity a variety receives, the more likely it will continue to be grown each year. That's how I came to grow Floriani Red Flint corn in 2013. *Mother Earth News* magazine had had articles[16] about it, and Southern Exposure Seed Exchange had it in their catalog and highly recommended it. I had also heard personal recommendations, so although I had been growing Bloody Butcher for more than two decades, I kept an open mind and tried it. The yield was smaller than I would have liked, but I found I like the taste of Floriani Red better than Bloody Butcher and am convinced that if I work with it, the yield will increase.

Become an Activist

Seed saving and sharing is a great thing, but you need to have a garden to grow seeds. If you are gardenless and still want to be involved, there

Mobile Seed Story Broadcasting Station. CREDIT: JEANETTE HART-MANN

are ways to do that. You can write, draw, and sing about seeds, which are activities I suggested to get a seed library off the ground or to keep one going. Let everyone know the plight of seeds and what they can do about it. You can be the one who records the stories of the seeds and the savers. Someone has to tell the stories — not only about specific varieties, but about the people and the culture.

You could retrofit an old bread truck into a solar-powered, grassroots roving, seed story shout-out vehicle committed to examining the interconnections between people and agriculture through performance, listening, and sharing of stories, resources, and seeds. That's exactly what SeedBroadcast[17] did when it put its Mobile Seed Story Broadcasting Station on the road touring the country in search of seedy partners at farmers markets, community gardens, and anywhere they could find them; they share the stories of the people with the world through the wonders of the Internet.[18] If you listen to the stories people have to tell about seeds, you will be informed and inspired about what seeds mean to them. SeedBroadcast developed from an idea shared by Jeanette Hart-Mann and Chrissie Orr to investigate food culture in action. The Mobile Seed Story Broadcasting Station is only one part of their work.

CHAPTER 11

We Are Living in Exciting Times!

S LOWLY (or not so slowly), our governments have let multinational corporations take control of our food, and thus our health and our lives. To make matters worse, we went along with it! Fortunately, more and more people are realizing that we can choose our own future. If that future means freedom from genetically altered seeds and crops, we have to change our ways to get them out of our lives. We each are responsible for any change we want to see and experience. The first step is to stop eating food that was grown with GMO seed. Since that includes much of what is offered in the grocery stores and served at restaurants, you will probably need to head to the farmers markets or step into your garden. You could also campaign for proper labeling of GMOs. That must be done so that consumers who are at the mercy of the present food system can make wise choices. However, now you know that you don't have to be in that position. With every bite you take, you are voting for how you want the Earth used to produce your food.

This book is about saving open pollinated seed to preserve the biodiversity of the planet, to preserve our culture, and to keep the seeds in the hands of the people. If you really believe that, you should be eating food grown from open pollinated seed. You can grow it yourself or you can find local growers at the farmers markets and other local food outlets. Ask the farmers the source of their seed and if what they grow is open pollinated and non-GMO. Don't be too surprised if they are growing hybrid varieties. They have to buy those from the seed companies each year, but they may be convinced that the yield is higher. If that is the case, maybe there should be a premium for what they grow from open pollinated varieties. Are you willing to pay a premium?

A hybrid variety may be bred for higher yields, but the plants can only do so much. If a plant puts out more pounds of food, that might mean that each pound has a little less of the nutrients than it would have if there were fewer pounds produced. With a higher yield, the nutrients gathered from the soil by the root system of each plant are divided among more pounds. It could also mean that the extra pounds are made up from more water. Flavor and nutrition both suffer in these scenarios. Did you ever read the fine-print on the packaging of your Thanksgiving turkey? Those self-basting turkeys have liquid pumped into them. The percentage of that is on the package. A pasture-raised organic turkey will cost more per pound, but there will be nothing pumped in, yielding more meat from the same weight as a conventionally raised bird. The organic turkey is more nutrient dense. With our fruits and vegetables, nutrient dense food is what we should be looking for.

The most important thing to keep in mind is that we have choices. We are living in exciting times because people are standing up to maintain those choices. For some of us who have been growing and saving open pollinated seeds and working to keep the seeds in the hands of the people for many years, the road we have been on has certainly been the one less traveled. I sometimes think of what I do as walking through the deep snow after a storm. Once there is a path, it is much easier for others to follow. However, I don't want people to follow me; I want them to walk with me. With the phenomenal growth in the number of seed libraries, there are many paths — highways actually — to make traveling this journey easier. We can all learn from each other. Join the social networks for seed libraries[1] so you can exchange ideas with others doing the same thing. No doubt, by the time you read this book there will be many more Internet resources for you to find. You can create your own face-to-face learning by visiting seed libraries in your area or in your travels to see how they operate. Maybe you will be the one to organize a gathering of seed librarians in your region.

Where Are They Now?

I couldn't help but wonder what had happened to those who opened up the first seed library path, so I did a little searching and found out more about the founders of BASIL and some other seed libraries I've told you about.

Sascha DuBrul is the young man I mentioned in Chapter 1 whose idea it was to start the first seed library in Berkeley (BASIL); then he moved to upstate New York and influenced someone to start one there. He had a case of wanderlust, a diagnosis of bipolar disorder, and a penchant for activism that kept him moving around for years. DuBrul is not satisfied with how mental health is handled by the establishment meant to deal with these things and feels that much more can be done for treatment besides passing out medication. To that end, he is the co-founder of The Icarus Project,[2] a radical community support network and media project that's actively redefining the language and culture of mental health and illness. He has written his autobiography, *Maps to the Other Side: The Adventures of a Bipolar Cartographer* and, through The Icarus Project, published *Navigating the Space Between Brilliance and Madness: A Reader and Roadmap of Bipolar Worlds.* He is one of the writers for Mad in America,[3] a website designed to serve as a resource and a community for those interested in rethinking psychiatric care in the US and abroad. DuBrul sees a mental condition that differs from the norm as a gift that needs to be understood. We are living in changing and exciting times. We need people who can think beyond the current conditions and come up with new ideas to lead our society to a peaceful, welcoming future. Just maybe, Sascha DuBrul and those like him will show us the way.

Christopher Shein had a hand in starting the seed library in Berkeley, along with DuBrul and Terri Compost. Shein now owns Wildheart Gardens, a permaculture landscape business, and teaches permaculture at Merritt College, one of the first community colleges in the nation to offer a certificate in Permaculture Design. His work at Merritt has resulted in an acre of food forest that he maintains with his students. Shein and his family grow food in their own backyard in the East Bay area of California. With help from Julie Thompson, he wrote *The Vegetable Gardener's Guide to Permaculture: Creating an Edible Ecosystem,* which was published by Timber Press in 2013.

Terri Compost kept the seed program alive for years at the Ecology Center. In 2009, her book *People's Park: Still Blooming* was published by Slingshot Collective. It tells the story of a park that developed as the result of radical activism in the late 1960s at the University of California at Berkeley and what has happened to it in the years since. It is a tale of citizens keeping the land in the commons for all to use. Compost provided assistance to Rebecca Newburn when she was starting the Richmond Grows Seed Lending Library in 2010. Terri Compost resides in Washington state now; working on a farm and still saving seeds.

Rebecca Newburn is the person who was instrumental in getting the Richmond Grows Seed Lending Library established and posted the instructions for others to follow. She really opened up that road. She continues to be an active volunteer with the seed library and is the one who adds your name to the Sister Seed Libraries list when you fill out the form at www.seedlibraries.net. Newburn, attended the very first Seed School in Cornville, Arizona, in 2010 and was involved with a three-day intensive Seed Library School at NS/S in 2012.

Setting up models for others to follow is Newburn's style. In cooperation with the Richmond Rivets, she has developed the *2 Steps a Month to Emergency Preparedness* program that she uses in her classroom as part of earth science education studying severe weather, plate tectonics, and climate change. The program helps individuals and communities organize themselves to be more prepared in case of emergencies. You can follow the program yourself from the information she has posted on the Internet.[4] In *Grow a Sustainable Diet,* I wrote that real living comes when everything you do becomes one joyful whole. Rebecca Newburn is certainly an example of that.

Amanda West Montgomery is the graduate student responsible for the seed library at the Lawrenceville branch of the Carnegie Library in

Pittsburgh, Pennsylvania. After graduation from Chatham University, she moved to Richmond, Virginia, and took a job at the Weinstein Jewish Community Center. She is the garden coordinator there and an assistant preschool teacher. The preschool teachers have a garden space for their classes, each with their own theme, such as pizza, salsa, dye, etc. Through Montgomery's involvement, the preschool classes take part in the Two Bite Club,[5] a USDA program encouraging children to eat healthy foods. (There is also a Dig In! program for grades 5–6 available from USDA.) Montgomery manages the garden component of the summer camps for 2-year-olds through 6th-grade children. She has found a way to combine her agriculture background with gardens and working with children.

Follow Your Heart

The people I've mentioned in this chapter and throughout the book have one thing in common besides seeds. They each followed their heart and acted accordingly. They dared to try out new ideas and open the paths for others to come with them. Starting a seed library is a leap of faith. You have to have faith that gardeners in your community will come forward and join you in this endeavor. You are building a network of sharing and caring. Some seed libraries have been planned that never really got off the ground. Some were started, but quietly closed after a couple years. It could be they are in limbo, waiting for this book to be published to help them jumpstart their project. Everything happens in its own time.

Follow your heart as you go forward nurturing your plants and saving the seeds. You have the opportunity to work in cooperation with Mother Nature and with seeds to secure a richer future that has nothing to do with money. We know what happened when corporations took control of seeds and followed the money. Is it any wonder that the health of the people and the health of the soil have declined?

People are coming to realize that we are living in a dynamic world. The things around us aren't static, but alive and changing all the time — just as seeds are. We have to find a way so that saving seeds for the future becomes as much of a way of life as planting seeds to grow. Every garden should have seeds as a crop for at least some of the harvest. People are longing for stability in these evolving times. Seeds can connect them with life and culture that has gone before and that is to come. You can be an active participant in making that happen. Yes, we are living in exciting times!

Afterword

SEED LIBRARIES ARE AN EVOLVING PHENOMENON. In fact, they are evolving faster than I can write about them. As soon as I finished writing this book, I learned of a new seed library in Pennsylvania and the challenges they are facing. The Pennsylvania Department of Agriculture (PA DOA) alerted the Simpson Seed Library[1] that they would be in violation of the state seed laws if they operated as planned. Their plans were to proceed as most seed libraries do — making seeds available that were grown and saved by patrons. According to the PA DOA's interpretation of the seed laws, the seed library would have to abide by the regulations in place that oversee seed companies. The fact that the patrons were receiving seeds for free didn't matter. The Simpson Seed Library decided to follow the recommendations of the PA DOA and only stock seeds commercially packaged for the current year. They encouraged their patrons to save seeds themselves and participate in a seed swap that the library would host. According to the PA DOA, seed *swaps* at the library were legal, but seed distribution by the library was not.

It is too bad that happened, but at the same time, things like that show that there needs to be more understanding about the issues involved. The PA DOA felt that any distribution of seeds should be regulated according to their seed laws, which were put in place to protect consumers from unscrupulous seed companies. But, seed libraries are not seed companies. Patrons of seed libraries participate in these seed sharing opportunities for all the reasons you have read about in this book. They know that the seeds at the seed library don't necessarily meet the legal germination rate, although often they could well exceed it. The amount of seed exchanged is so small that if there were any unwanted seeds present, such as noxious weeds, those seeds could be easily identified and expelled from the batch. The chance that seed is not true-to-type is greater if acquired from amateur seed savers than if bought from a reputable seed company. That's the chance you take when you get free seeds from a seed library. Our seed laws help protect us from seeds with low germination, noxious weeds, and seeds that are not as advertised — when purchased from seed companies.

With seed libraries, people are looking for something different than what is available from seed companies. They are looking for a personal connection with the seed savers and with varieties specific to their area. Those in the seed library world are wondering if other seed libraries will face similar challenges. I imagine that if they do, the issue will eventually end up in court. The best defense is to promote education on the matter of seed libraries, not just to the seed savers, but to the general public. Let them know the value of local seeds. You have the opportunity to put that message forward in so many ways using the ideas I've presented in this book. Changes may need to be made in the seed laws. All states have seed laws, but they are not worded the same way. It could come down to semantics — how the laws are worded — and interpretation of that wording. With that in mind, it has been suggested[2] that your seed library not have patrons enter into contracts, such as signing a form saying they will bring back seeds. It may be best to only have them register to exchange information, such as their contact information, the seeds they take and donate, and their experiences with the seeds.

The Simpson Seed Library will host a seed swap in their library. Seed swaps are great. In fact, I think that if a public library was considering starting a seed library, but was hesitant to delve into storing and packaging the seeds themselves, hosting regular seed swaps would be the way to go. A seed swap could be held every time there is a program relating to seeds or gardening. It is a great approach to ease into the seed world. The library would not be responsible for the seeds themselves, but could go about all the other activities suggested in this book. The spot that may have been designated for seeds could still be set aside for a seed share program. It could have related resources available, along with an opportunity for patrons to leave recipes and ideas to share with others. It may also be the spot to keep seed-related equipment that is available for loan. Seed screens come to mind for that.

We are living in ever-changing times. Before this book is released, seed libraries may be in the news again about issues we haven't even thought of yet. If you have joined the seed networks I've mentioned, you will be kept abreast of the latest developments. Join the fun and know that you are part of something good.

Resources

Eating Locally

Halweil, Brian, *Eat Here: Reclaiming Homegrown Pleasures in a Global Supermarket.* New York: W.W. Norton & Company, 2004.

Kingsolver, Barbara, *Animal, Vegetable, Miracle: A Year of Food Life.* New York: Harper Collins, 2007.

Nabhan, Gary Paul, *Coming Home to Eat: The Pleasures and Politics of Local Foods.* New York: W.W. Norton & Company, 2002.

Robin, Vicki, *Blessing the Hands That Feed Us: What Eating Closer to Home Can Teach Us About Food, Community, and Our Place on Earth.* New York: Viking Penguin, 2014.

Smith, Alisa and J.B. Mackinnon, *Plenty: One Man, One Woman, and a Raucous Year of Eating Locally.* New York: Harmony Books, 2007. Also published as *The 100-Mile Diet: A Year of Local Eating.*

Gardening

Coleman, Eliot, *Four-Season Harvest: Organic Vegetables from Your Home Garden All Year Long.* rev. ed., White River Junction, VT: Chelsea Green, 1999.

Conner, Cindy, *Cover Crops and Compost Crops IN Your Garden.* DVD, Ashland, VA: Homeplace Earth, 2008.

Conner, Cindy, *Develop a Sustainable Vegetable Garden Plan*. DVD, Ashland, VA: Homeplace Earth, 2010.

Conner, Cindy, *Grow a Sustainable Diet: Planning and Growing to Feed Ourselves and the Earth*. Gabriola Island, BC: New Society Publishers, 2014.

Damrosch, Barbara and Eliot Coleman, *The Four Season Farm Gardener's Cookbook: From the Garden to the Table in 120 Recipes*. New York: Workman, 2012. Contains plans for making and using a 10′ x 12′ portable greenhouse.

Deppe, Carol, *The Resilient Gardener: Food Production and Self-reliance in Uncertain Times*. White River Junction, VT: Chelsea Green, 2010.

Dawling, Pam, *Sustainable Market Farming: Intensive Vegetable Production on a Few Acres*. Gabriola Island, BC: New Society Publishers, 2013.

Gilkeson, Linda, *Backyard Bounty: The Complete Guide to Year-round Gardening in the Pacific Northwest*. Gabriola Island, BC: New Society Publishers, 2011.

Jeavons, John, *How to Grow More Vegetables and Fruits, Nuts, Berries, Grains, and Other Crops Than You Ever Thought Possible on Less Land Than You Can Imagine*, 8th ed. Berkeley, CA: Ten Speed Press, 2012.

Nabhan, Gary Paul, *Growing Food in a Hotter, Drier Land: Lessons from Desert Farmers on Adapting to Climate Uncertainty*. White River Junction, VT: Chelsea Green, 2013.

Shein, Christopher with Julie Thompson, *The Vegetable Gardeners Guide to Permaculture: Creating an Edible Ecosystem*. Portland, OR: Timber Press, 2013.

Timber Press Regional Gardening Guides, Portland, OR: Timber Press, 2013:

Forkner, Lorene Edwards, *The Timber Press Guide to Vegetable Gardening in the Pacific Northwest*.

Iannotti, Marie, *The Timber Press Guide to Vegetable Gardening in the Northeast*.

Newcomer, Mary Ann, *The Timber Press Guide to Vegetable Gardening in the Mountain States*.

Wallace, Ira, *The Timber Press Guide to Vegetable Gardening in the Southeast*.

Plant Breeding

Deppe, Carol, *Breed Your Own Vegetable Varieties: The Gardener's and Farmer's Guide to Plant Breeding and Seed Saving.* White River Junction, VT: Chelsea Green, 2000.

Tychonievich, Joseph, *Plant Breeding for the Home Gardener: How to Create Unique Vegetables and Flowers.* Portland, OR: Timber Press, 2013.

White, Rowen and Bryan Connelly, *Breeding Organic Vegetables: A Step-by-step Guide for Growers.* Rochester, NY: Northeast Organic Farming Association of New York, 2011.

Safe Seed Pledge

Council for Responsible Genetics Safe Seed Resource List www.council forresponsiblegenetics.org/ViewPage.aspx?pageId=261

Seed Banks

Seeds of Diversity Canada, *Micro-Seedbanking: A Primer on Setting Up and Running a Community Seed Bank.* www.seeds.ca/int/doc/docpub.php?n=web/sl/doc/Micro-Seedbanking-primer.pdf

Seed Inventories

Seed Savers Exchange, *Garden Seed Inventory,* 6th ed. 2004. An inventory of non-hybrid vegetable varieties from mail order vegetable seed companies in the US and Canada. 503-page book.

Seeds of Diversity Canada, *Canadian Seed Catalogue Inventory.* Online list of vegetable and fruit seeds that were sold in recent years by Canadian Seed Companies. www.seeds.ca/sl/csci/

Seed Library Resources

www.seedlibraries.net Website maintained by Rebecca Newburn with information about how to set up a seed library. Here you will find the Sister Seed Libraries list and seed librarians can exchange information.

www.seedlibrarics.org The Seed Library Social Network.

www.seedmatters.org Much about seeds here. You will find the Community Seed Resource Program helpful.

Seed Saving

Ashworth, Suzanne, *Seed to Seed: Seed Saving and Growing Techniques for Vegetable Gardeners,* 2nd ed. Decorah, IA: Seed Savers Exchange, 2002.

Bubel, Nancy, *The New Seed Starters Handbook.* Emmaus, PA: Rodale Press, 1988.

Gough, Robert E. and Cheryl Moore-Gough, *The Complete Guide to Saving Seeds: 322 Vegetables, Herbs, Fruits, Flowers, Trees, and Shrubs.* North Adams, MA: Storey, 2011.

Heistinger, Andrea, *The Manual of Seed Saving: Harvesting, Storing, and Sowing Techniques for Vegetables, Herbs, and Fruits.* Portland, OR: Timber Press, 2010.

Jason, Dan, *Saving Seeds As If Our Lives Depended on It,* 6th ed. Canada: Salt Spring Seeds, 2014.

Jeffery, Josie, *Seedswap: The Gardeners Guide to Saving and Swapping Seeds.* East Sussex, UK: Leaping Hare Press, 2012.

McCormack, Jeffery H., *Seed Processing and Storage: Principles and Practices of Seed Harvesting, Processing, and Storage.* An organic seed production manual for seed growers in the Mid-Atlantic and Southern US, 2004 www.carolinafarmstewards.org/wp-content/uploads/2012/05/SeedProcessingandStorageVer_1pt3.pdf

McDorman, Bill, *Basic Seed Saving.* Cornville, AZ: Seeds Trust, 1994. Online information available at www.seedsave.org/index.php/seed-saving-instructions

Organic Seed Alliance, *A Seed Saving Guide for Gardeners and Farmers.* www.seedalliance.org/uploads/publications/Seed_Saving_Guide.pdf

Organic seed production manuals specific to the Mid-Atlantic and South and to the Pacific Northwest are available at www.savingourseeds.org

Navazio, John, *The Organic Seed Grower: A Farmer's Guide to Vegetable Seed Production.* White River Junction, VT: Chelsea Green, 2012.

Rogers, Marc, *Saving Seeds: The Gardener's Guide to Growing and Storing Vegetables and Flower Seeds.* North Adams, MA: Storey, 1990.

Seeds of Diversity Canada, *How to Save Your Own Seeds: A Handbook for Small-Scale Seed Production.* Available in both English and French editions from Seeds of Diversity Canada. www.seeds.ca

Strickland, Sue, *Back Garden Seed Saving: Keeping Our Vegetable Heritage Alive*. Bath, UK: Eco-Logic Books, 2008.

Turner, Carol B., *Seed Sowing and Saving: Step-by-step Techniques for Collecting and Growing More Than 100 Vegetables, Flowers, and Herbs*. North Adams, MA: Storey, 1998.

Seed Saving with Children

Kaufman, Eli Rogosa, *An Activity Guidebook in the Living Tradition of Seed Saving*. Waterville, ME: Fedco Seeds, 2001. www.fedcoseeds. com/forms/seedschool.pdf

Seed Saving History and People

Ausubel, Kenny, *Seeds of Change: The Living Treasure*. San Francisco, CA: Harper San Francisco, 1994.

Best, Bill, *Saving Seeds, Preserving Taste: Heirloom Seed Savers in Appalachia*. Athens, OH: Ohio University Press, 2013.

Nabhan, Gary Paul, *Enduring Seeds: Native American Agriculture and Wild Plant Conservation*. Tucson, AZ: University of Arizona Press, 1989.

Newman, Judy, ed., *Every Seed Tells a Tale: Stories of Plants, People, and Places That Have Contributed to Canada's Seed Heritage*. Canada: Seeds of Diversity Canada, 2009.

Ray, Janisse, *The Seed Underground: A Growing Revolution to Save Food*. White River Junction, VT: Chelsea Green, 2012.

Rempel, Sharon, *Demeter's Wheats: Growing Local Food and Community with Traditional Wisdom and Heritage Wheat*. Canada: Grassroot Solutions, 2008.

Whealy, Diane Ott, *Gathering: Memoir of a Seed Saver*. Decorah, IA: Seed Savers Exchange, 2011.

Wilson, Gilbert L., *Buffalo Birdwoman's Garden*. St Paul, MN: Minnesota Historical Society Press, 1987. Originally published as *Agriculture of the Hidatsa Indians: An American Interpretation* in 1917.

Seed Saving Organizations

Heritage Seed Library www.gardenorganic.org.uk/hsl Saving heritage seed varieties in the UK.

Irish Seed SaversAssociation www.irishseedsavers.ie Seed saving in Ireland.

Native Seeds/SEARCH www.nativeseeds.org Administration, seed bank, and mail order: 3584 E. River Road Tucson, AZ 85718; Retail store: 3061 N. Campbell Avenue, Tucson, AZ 85719. Phone 520-622-0830.

Navdanya www.navdanya.org Vandana Shiva's organization in India.

Organic Seed Alliance www.seedalliance.org P.O. Box 772, Port Townsend, WA 98368. Phone 360-385-7192.

Rocky Mountain Seed Alliance www.rockymountainseeds.org Box 4736, Ketchum, ID 83340. Phone 928-300-7989.

Seed Savers Exchange www.seedsavers.org 3094 North Winn Road, Decorah, IO 52101. Phone 563-382-5990.
Find information about the Community Seed Resource Program (CSRP) at www.exchange.seedsavers.org/csrp/about.aspx CSRP is a collaboration between Seed Savers Exchange and Seed Matters.

Seeds of Diversity Canada www.seeds.ca P.O. Box 36, Stn Q, Toronto, ON M4T 2L7. Phone 866-509-7333.

Seed Saving Supplies

Bountiful Gardens www.bountifulgardens.org 1712-D South Main St., Willits, CA 95490. Phone 707-459-6410.

Horizon Herbs www.horizonherbs.com P.O. Box 69, Williams, OR 97544. Phone 541-846-6704. Seed cleaning screens.

Seed Savers Exchange www.seedsavers.org 3094 North Winn Road, Decorah, IO 52101. Phone 563-382-5990.

Southern Exposure Seed Exchange www.southernexposure.com P.O. Box 460, Mineral, VA 23117. Phone 540-894-9480.

Sustainable Seed Company www.sustainableseedco.com P.O. Box 38, Cavelo, CA 95428. Phone 877-620-7333.

Territorial Seed Company www.territorialseed.com 20 Palmer Ave., Cottage Grove, OR 97424. Phone 800-626-0866.

Saving Our Seeds www.savingourseeds.org/pdf/basic_seed_cleaning_equipment.pdf List of equipment needed for seed growers and seed savers.

Seed Schools

Hawai'i Public Seed Initiative www.kohalacenter.org/hpsi The Kohala Center, 65-1291A Kawaihae Road, Kamuela, HI 96743.

Rocky Mountain Seed Alliance www.rockymountainseeds.org See "Seed Saving Organizations."

Seed Academy at Seven Seeds Farm sevenseedsfarm.com, 3220 E. Fork Road, Williams, OR 97544. Phone 541-846-9233.

Seed Swaps

How to Host a Seed Swap www.southernexposure.com/how-to-host-a-seed-swap-ezp-146.html

TED Talks

Jonathan Drori, "Why We're Storing Billions of Seeds," 2009. www.ted.com/talks/jonathan_drori_why_we_re_storing_billions_of_seeds

Cary Fowler, "One Seed at a Time, Protecting the Future of Food," 2009. www.ted.com/talks/cary_fowler_one_seed_at_a_time_protecting_the_future_of_food#t-455064

Winona LaDuke, "Seeds of Our Ancestors, Seeds of Life," 2011. www.tedxtalks.ted.com/video/TEDxTC-Winona-LaDuke-Seeds-of-O;search%3Atag%3A%22TEDxTC%22

Bill McDorman, "Seeds Will Save Us," 2013. www.youtube.com/watch?v=pWDsDRQJHtg&feature=kp McDorman talks about seed libraries at the end of the talk — good introduction to why you want to save seeds.

Simran Sethi, "Seeds: the buried beginnings of food," 2013. www.tedxtalks.ted.com/video/Seeds-The-Buried-Beginnings-of

Vandana Shiva, "Solutions to the Food and Ecological Crisis Facing Us Today," 2010. www.youtube.com/watch?v=ER5ZZk5atlE

Videos

Garcia, Deborah Koons, *The Future of Food.* Mill Valley, CA: Lily Films, 2008.

Joanes, Sofia, *Fresh: The Movie.* Milwaukee, WI: Ripple Effect Films, 2009.

Kenner, Robert, *Food, Inc.* New York: Magnolia Pictures, 2009.

Morgan, Faith, Eugene Murphy, and Megan Quinn, *The Power of Community: How Cuba Survived Peak Oil.* Yellow Springs, OH: Community Services, Inc., 2006.

Kaminsky, Sean, *Open Sesame: The Story of Seeds.* 2014. www.opensesamemovie.com

Smith, Miranda, Abigail Wright, and Nathaniel Kahn, *My Father's Garden.* Oley, PA: Bullfrog Films, 1995.

Tomei, Marisa, executive producer with Taggart Siegel and John Betz, filmmakers, *Seed: The Untold Story.* Portland, OR: Collective Eye Films, 2015.

Webinars and Other Broadcasts

How to Start a Seed Library webinar by The Center for a New American Dream. www.newdream.org/blog/seed-library

Seed Savers Exchange has free monthly webinars at www.seedsavers.org/Education/Webinars/ You can subscribe to stay informed on dates, times, and registration details.

Hear seed stories gathered by the Mobile Seed Story Broadcasting Station at www.soundcloud.com/seedbroadcast Visit the website at www.seedbroadcast.org

Notes

Chapter 1

1. This information comes from the ETC Group (Action Group on Erosion, Technology, and Concentration), which monitors the impact of emerging technologies and corporate strategies on biodiversity, agriculture, and human rights. See: etcgroup.org/content/just-3-companies-control-more-half-53-global-commercial-market-seed

2. Jack Doyle, *Altered Harvest: Agriculture, Genetics, and the Fate of the World's Food Supply.* New York: Penguin Books,1986, pp. 97–99.

3. Online article "Monsanto Decimates their Credibility" by Dr. Mercola, September 10, 2013. articles.mercola.com/sites/articles/archive/2013/09/10/monsanto-bt-corn.aspx

4. A chart comparing the nutritional value of GMO corn to non-GMO corn in 2012 can be found at: www.dedellseeds.com/media/pdf/Comparison_of_GMO_Corn_versus_Non-GMO_Corn_REV1.pdf

5. *Navdanya: Two Decades of Service to the Earth and Small Farmers,* a collection of articles about Navdanya and Vandana Shiva: www.navdanya.org/attachments/Navdanya.pdf

6. Janice M Strachan, "Plant Variety Protection: An Alternative to Patents," www.nal.usda.gov/pgdic/Probe/v2n2/plant.html Published in *Probe* Volume 2(2): Summer 1992.

7. Plant Variety Protection Office Certificate Status Database: www.ars-grin.gov/cgi-bin/npgs/html/pvplist.pl

8. Definitions of Utility Patent and PVP can be found at: www.johnnyseeds.com/assets/information/understanding_utility_patents_and_pvp.pdf

9. Information about plant patents from the United States Patent and Trademark Office can be found at: www.uspto.gov/web/offices/pac/plant/#1

10. Bill McDorman and Stephen Thomas, "Sowing Revolution: Seed Libraries Offer Hope for Freedom of Food," *Acres USA* January 2013, pp. 20–26. www.nativeseeds.org/pdf/Jan2012_McDorman Thomas.pdf

Chapter 2

1. Susan Campbell Bartoletti, *Black Potatoes: The Story of the Great Irish Famine, 1845-1850.* Boston: Houghton Mifflin Company, 2001.

2. Jack Doyle, *Altered Harvest,* p. 13.

3. Michael Pollan, *The Omnivore's Dilemma: A Natural History of Four Meals.* New York: Penguin Press, 2006.

4. Kent Whealy, original editor, updated by Joanne Thuente, *Garden Seed Inventory.* 6th ed. Decorah, IA: Seed Savers Exchange, 2004.

5. Jerry Minnich, *Gardening for Maximum Nutrition: Easy Ways to Double the Nutritional Value of Your Backyard Garden.* Emmaus, PA: Rodale Press, 1983, p. 4.

6. Cindy Conner, *Grow a Sustainable Diet: Planning and Growing to Feed Ourselves and the Earth.* Gabriola Island, BC: New Society Publishers, 2014.

Chapter 4

1. The official website of the Transition Network is www.transitionnet work.org Rob Hopkins' blog, Transition Culture, is at www.transition network.org/blogs/rob-hopkins

2. The official website of the Transition Network for the United States is www.transitionus.org A map showing Transition Initiatives worldwide can be found at www.transitionnetwork.org/initiatives/map

3. Find an informative video about Permaculture GTA here: www.permaculturegta.org

4. Lopez Community Land Trust Seed Library: www.lopezclt.org/seed-security-initiative-and-seed-library/

5. Hull House Museum Seed Library: www.uic.edu/jaddams/hull/_programsevents/_kitchen/_seedlibrary/seedlibrary.html

6. Canada Master Gardeners: www.ahs.org/gardening-resources/master-gardeners/canada-master-gardeners

7 UK Master Gardeners: mastergardeners.org.uk/

Chapter 5

1. Gilbert L. Wilson, *Buffalo Bird Woman's Garden.* St. Paul, MN: Minnesota Historical Society Press, 1987. Originally published as *Agriculture of the Hidatsa Indians: An American Interpretation,* in 1917.

2. Council for Responsible Genetics Safe Seed Resource List: www.councilforresponsiblegenetics.org/ViewPage.aspx?pageId=261

3. Jeffrey H. McCormack, PhD, "Seed Processing and Storage," version 1.3, December 28, 2004, p. 9.

4. Ibid.

5. Seed saving resources, including a seed saving chart, provided by Seed Matters and Seed Savers Exchange, exchange.seedsavers.org/csrp/resources.aspx

6. John Jeavons, *How to Grow More Vegetables and Fruits, Nuts, Berries, Grains, and Other Crops.* 8th ed. New York, NY: Ten Speed Press, 2012.

7. Carol Deppe, *Breed Your Own Vegetable Varieties.* White River Junction, VT: Chelsea Green, 2000, pp. 232–236.

8. Website for Seed Savers Exchange: www.seedsavers.org

9. Heritage Seed Library in the UK: www.gardenorganic.org.uk/hsl/

10. Irish Seed Savers Association: www.irishseedsavers.ie/

11. Seed Savers Network in Australia: seedsavers.net/
12. Seed School at Rocky Mountain Seed Alliance: rockymountainseeds.org/attend/seed-school
13. Rocky Mountain Seed Alliance: rockymountainseeds.org/
14. David King, "Why Seed School," SLOLA blog, July 27, 2013. slola.blogspot.com/2013/07/why-seed-school.html

Chapter 6

1. Concord Seed Lending Library: www.concordseedlendinglibrary. org/
2. Brochure for the Pima County Seed Library: www.library.pima. gov/seed-library/seed_library_saving.pdf

Chapter 7

1. The Organic Seed Alliance. "A Seed Saving Guide for Gardeners and Farmers." Port Townsend, WA, 2010: www.seedalliance.org/ uploads/publications/Seed_Saving_Guide.pdf
2. Saving Our Seeds: www.savingourseeds.org/publications.html
3. *How to Start a Seed Library at Your Public Library* webinar www. newdream.org/resources/webinars/seed-library

Chapter 9

1. Smith, Alissa, "Fairfield Preschoolers Help Local Seed Library Grow," *Fairfield Daily Voice,* March 3, 2014. fairfield.dailyvoice. com/neighbors/fairfield-preschoolers-help-local-seed-library-grow

Chapter 10

1. Portland Seed Library, Portland, Oregon: portlandseedlibrary.net/
2. Seeds of Diversity link to Seedy Day information: www.seeds.ca/ int/doc/docpub.php?n=web/ebulletin/other/organizing-a-seedy-saturday
3. Information about Seedy Day events in the UK: seedysunday.org. uk
4. Information about Seedy Days in the US: seedswapday.blogspot. com

5. Svalbard Global Seed Vault: www.atlasobscura.com/places/svalbard-seed-bank

6. Salt Spring Seed and Plant Sanctuary: www.seedsanctuary.com/

7. Jason, Dan, *Saving Seeds As If Our Lives Depended on It.* 6th ed. Salt Spring, BC, Canada: Salt Spring Seeds, 2014, p. 47.

8. Rainwater, Ellen, "How to Start a Community Seed Bank," *Permaculture Activist* Spring 2014, pp. 35–36.

9. Navdanya website: www.navdanya.org/about-us/mission

10. Video of Vandana Shiva giving her "Seeds of Love" talk: www. Seedfreedom.in

11. Seed Share station plans: nomoola.com/seeds/diy.html

12. Dawling, Pam, *Sustainable Market Farming: Intensive Vegetable Production on a Few Acres.* Gabriola Island, BC: New Society Publishers, 2013.

13. Ira Wallace is author of *The Timber Press Guide to Vegetable Gardening in the Southeast.* Portland, OR: Timber Press, 2013.

14. Wildfong, Bob, "Happy Birthday, Seeds of Diversity!" *Seeds of Diversity* Winter 2014, p. 4.

15. Root and Radicle Seed Company: www.rootandradicleseedco.com/

16. Rubel, William, "The Return of a Great Corn Variety," *Mother Earth News* February/March 2009; and "Floriani Red Flint Corn: The Perfect Staple Crop," *Mother Earth News* December 2010/January 2011.

17. Learn more about SeedBroadcast at seedbroadcast.org

18. Listen to audio broadcasts of seed stories at soundcloud.com/seedbroadcast

Chapter 11

1. Seed Libraries Social Network: www.seedlibraries.org/ Find a seed library forum at www.seedlibraries.net

2. The Icarus Project: www.theicarusproject.net

3. Mad in America: www.madinamerica.com

4. 2 Steps a Month to Emergency Preparedness: 2stepsamonth.word press.com/about

5. The Two Bite Club: www.fns.usda.gov/tn/two-bite-club

Afterword

1. The Simpson Seed Library website, with an explanation about how they are working with the Pennsylvania Department of Agriculture: www.cumberlandcountylibraries.org/?q=SIM_SeedLibrary

2. An article about the legality of seed libraries written by Janelle Orsi and Neil Thapar of the Sustainable Economies Law Center, with input from Neal Gorenflo of Shareable and Sarah Baird of the Center for a New American Dream: www.shareable.net/blog/setting-the-record-straight-on-the-legality-of-seed-libraries

Index

About the Author

Cindy Conner lives near Ashland, Virginia with her husband on their five-acre farm. Raising four children there, being a market gardener for ten years and teaching sustainable agriculture at the local community college for over a decade kept her busy as she delved deeper into the sustainability of her actions. She produced two videos — *Cover Crops and Compost Crops IN Your Garden and Develop a Sustainable Vegetable Garden Plan* — as teaching tools for individuals or groups. These DVDs are still part of the college curriculum where she taught. Her book *Grow a Sustainable Diet: Planning and Growing to Feed Ourselves and the Earth,* published in 2014, guides the reader through planning a diet to grow that would produce a substantial part of one's food and keep a small footprint on the earth, while also growing cover crops to feed the soil.

Besides researching how to grow a complete diet in a small space, Cindy Conner has fun with other projects, such as growing and spinning cotton from her garden. One day she will exchange the quilted vest she wears with one made from homegrown, handspun, handwoven cotton. Watch for her as a presenter at the *Mother Earth News* Fairs. You can follow Cindy's blog atHomplaceEarth.wordpress.com.

If you have enjoyed *Seed Libraries* you might also enjoy other

BOOKS TO BUILD A NEW SOCIETY

Our books provide positive solutions for people who want to make a difference. We specialize in:

Sustainable Living • Green Building • Peak Oil
Renewable Energy • Environment & Economy
Natural Building & Appropriate Technology
Progressive Leadership • Resistance and Community
Educational & Parenting Resources

New Society Publishers

ENVIRONMENTAL BENEFITS STATEMENT

New Society Publishers has chosen to produce this book on recycled paper made with **100% post consumer waste,** processed chlorine free, and old growth free.

For every 5,000 books printed, New Society saves the following resources:[1]

20	Trees
1,833	Pounds of Solid Waste
2,016	Gallons of Water
2,630	Kilowatt Hours of Electricity
3,332	Pounds of Greenhouse Gases
14	Pounds of HAPs, VOCs, and AOX Combined
5	Cubic Yards of Landfill Space

[1]Environmental benefits are calculated based on research done by the Environmental Defense Fund and other members of the Paper Task Force who study the environmental impacts of the paper industry.

For a full list of NSP's titles, please call 1-800-567-6772 *or check out our website* at:

www.newsociety.com